MATHEMATICAL PROGRAMS FOR ACTIVITY ANALYSIS

This volume was produced in collaboration with
Leuven University Press

MATHEMATICAL PROGRAMS FOR ACTIVITY ANALYSIS

Editor

PAUL VAN MOESEKE

Professor of Economics
University of Louvain

1974

NORTH-HOLLAND PUBLISHING COMPANY - AMSTERDAM • LONDON
AMERICAN ELSEVIER PUBLISHING CO., INC. - NEW YORK

Library of Congress Catalog Card Number: 74-75580
ISBN North-Holland 07204 28009
ISBN American Elsevier 0444 106332

PUBLISHERS:

NORTH-HOLLAND PUBLISHING COMPANY — AMSTERDAM
NORTH-HOLLAND PUBLISHING COMPANY, LTD.—LONDON

SOLE DISTRIBUTORS FOR THE U.S.A. AND CANADA:

AMERICAN ELSEVIER PUBLISHING COMPANY, INC.
52 VANDERBILT AVENUE
NEW YORK, N.Y. 10017

PRINTED IN BELGIUM

CONTRIBUTORS TO THIS VOLUME

MORDECAI AVRIEL, Technion, Haifa
TAMAR DAR, Tel Aviv University
GUY T. de GHELLINCK, University of Louvain and CORE
PIERRE DOULLIEZ, Société de Traction et d'Électricité, Brussels, and University of Louvain
IRINEL DRAGAN, University of Jassy
JACQUES H. DRÈZE, University of Louvain and CORE
JACQUES FEDERWISCH, SORCA and Institut des Hautes Études Commerciales, Brussels
PIERRE HANSEN, Institut d'Économie Scientifique et de Gestion, Lille
E. JAMOULLE, Société de Traction et d'Électricité, Brussels
LEON KAUFMAN, University of Brussels
ETIENNE LOUTE, University of Louvain and CORE
MAURICE G. MARCHAND, University of Louvain and CORE
URY PASSY, Technion, Haifa
ASSAF RAZIN, Tel Aviv University
JEAN-FRANÇOIS RICHARD, University of Louvain and CORE
M. RIJCKAERT, University of Louvain
M. ROUBENS, Faculté Polytechnique de Mons
MICHAEL RUBINOVITCH, Technion, Haifa
JOZEF L. TEUGELS, University of Louvain and CORE
PAUL VAN MOESEKE, University of Louvain and CORE
FRANS VAN WINCKEL, University of Louvain and CORE
KATRIEL WILD, IBM Israel
LAURENCE A. WOLSEY, University of Louvain and CORE
JOSEPH A. YAHAV, Tel Aviv University
ISRAEL ZANG, Technion, Haifa

PAPERS PRESENTED AT A PREVIOUS BELGO-ISRAELI COLLOQUIUM (NOT INCORPORATED IN THIS VOLUME)

STOCHASTIC SYSTEMS

Linear programming and Markov decision processes
GUY T. de GHELLINCK, University of Louvain and CORE
Probability bounds in replacement policies for Markov systems
AVRAHAM BEJA, Tel Aviv University
Application of Markov chains to the design of elevators
YORAM SEBBA, Technion, Haifa
Bounds for measures of performance of a GG1 & IFR G1 queuing system
Y. KINBERG, Israel Shipping Research Institute and I. SHLIFER, Technion, Haifa
Some stochastic models in reliability
BENJAMIN EPSTEIN, Technion, Haifa

MATHEMATICAL PROGRAMMING

Unconstrained optimization on the integers
BRUCE MILLER, The RAND Corporation and CORE
Reduction of linear constraints
ADAM SHEFI, Israeli Ministry of Defense
Complementary geometric programming
MORDECAI AVRIEL, Technion, Haifa
A deep-cut procedure for 0-1 linear programs
M. A. POLLATSCHEK and B. AVI-ITZHAK, Technion, Haifa
Application of 0-1 programming to outdoor recreation
MOSHE HILL and MORDECAI SHECHTER, Technion, Haifa
Progress in pseudo-Boolean programming
PETER L. HAMMER, Technion, Haifa
New algorithm for assignment of facilities to locations
Z. DREZNER, Israeli Ministry of Defense
Parametric flow in multiterminal networks and the capacity problem
PIERRE DOULLIEZ, University of Louvain and CORE

ECONOMIC APPLICATIONS OF OPERATIONS RESEARCH

Continuous-time model for the cash-balance problem
JEAN-PHILIPPE VIAL, University of Louvain and CORE
Capital adjustment with a U-shaped average cost of investment
YOAV KISLEV and HANA LIFSON
Homogeneous programming of efficient portfolios
PAUL VAN MOESEKE, University of Louvain and Iowa State University
Portfolio selection under uncertainty: a game-theoretic approach
A. BEN TAL and E. HOCHMAN, Tel Aviv Universit

Sed quemadmodum homines initio innatis instrumentis quaedam facillima, quamvis laboriose, & imperfecte, facere quiverunt, iisque confectis alia difficiliora minori labore, & perfectius confecerunt, & sic gradatim ab operibus simplicissimis ad instrumenta, & ab instrumentis ad alia opera, & instrumenta pergendo, eo pervenerunt, ut tot, & tam difficilia parvo labore perficiant; sic etiam intellectus vi sua nativa facit sibi instrumenta intellectualia, quibus alias vires acquirit ad alia opera intellectualia, & ex iis operibus alia instrumenta, seu potestatem ulterius investigandi, & sic gradatim pergit, donec sapientiae culmen attingat.

SPINOZA. Tractatus de Intellectus Emendatione

... c'est dans les siècles les plus barbares que se sont faites les plus utiles découvertes; il semble que le partage des temps les plus éclairés et des compagnies les plus savantes soit de raisonner sur ce que des ignorants ont inventé. On sait aujourd'hui, après les longues disputes de M. Huyghens et de M. Renaud, la détermination de l'angle le plus avantageux d'un gouvernail de vaisseau avec la quille; mais Christophe Colomb avait découvert l'Amérique sans rien soupçonner de cet angle.
Je suis bien loin d'inférer de là qu'il faille s'en tenir seulement à une pratique aveugle; mais il serait heureux que les physiciens et les géomètres joignissent, autant qu'il est possible, la pratique à la spéculation. ... Un homme, avec les quatre règles d'arithmétique et du bon sens, devient un grand négociant, ... tandis qu'un pauvre algébriste passe sa vie à chercher dans les nombres des rapports et des propriétés étonnantes, mais sans usage, et qui ne lui apprendront pas ce que c'est que le change. Tous les arts sont à peu près dans ce cas; il y a un point, passé lequel les recherches ne sont plus que pour la curiosité : ces vérités ingénieuses et inutiles ressemblent à des étoiles qui, placées trop loin de nous, ne nous donnent point de clarté.

VOLTAIRE. Lettres Philosophiques

PREFACE

This volume reports on the Belgo-Israeli Colloquium on Operations Research, August 29–31, 1972, held at the Pope Hadrian VI College, University of Louvain, where Israeli specialists on the one hand, and operations researchers of several nationalities working in Belgium on the other, communicated recent results in fields of current interest to both groups, scilicet: theory and algorithms of general, geometric, and integer programming, stochastic systems, networks, assignment problems, and sequential planning.

Most of the participants in this triduum had met at a previous Belgo-Israeli colloquium in Netanya, Israel, June 17–19, 1970, and comparison of the list of papers presented there (supra) with the Table of Contents (infra) indicates that several results collected in the following pages emanate from, or build on, investigations under mutual discussion at least since the Netanya meeting.

A series of similar conferences on mathematical programming were sponsored during the fifties and sixties by, *inter al.*, the RAND Corporation, the National Bureau of Standards, the US Air Force, and the Special Interest Group for Mathematical Programming (SIGMAP) within the Association for Computing Machinery. It is hoped that the present volume will take its place among the reports of these symposia, which have marked the growth of mathematical programming since its inception in 1947 by gathering in print the results of recent exchanges among groups of research workers on a number of *capita selecta*.

Whether this book marks one more modest step in Spinoza's gradation towards the perfection of scientific tools, without transgressing Voltaire's canon of relevance – concepts alluded to in the foregoing analects – is left for the reader to decide.

P.V.M.

ACKNOWLEDGMENTS

The Belgo-Israeli Colloquia were sponsored by the Operations Research Society of Israel (ORSIS) in collaboration with the Technion, Haifa, and by the Direction d'Administration des Relations Culturelles of the Belgian Foreign Office in cooperation with the Center for Operations Research and Econometrics (CORE) and the University of Louvain. Dean B. Avi-Itzhak of the Technion and Professor Jean-Philippe Vial of Louvain chaired the organization committees. The opening sessions of the Netanya and Louvain conferences were attended by the Belgian Ambassador in Jerusalem and the Israeli Ambassador in Brussels, respectively.

Thanks are due to Mr. G. Declercq, Administrator-General of Louvain University, who advised on the production of the manuscript; and to Drs. A. Beliën and Dr. A. Sümer of the Centrum voor Economische Studiën, who attended to most of the early editoıial work and correspondence.

I especially wish to thank Professors P. De Somer, Rector, and M. Eyskens, Commissar-General of Louvain University, who intervened to allow publication of this volume by North-Holland.

For their patience and cooperation I am indebted to Drs. W.H. Wimmers and Mr. E. Fredriksson, respectively Director and Editor at North-Holland; and to Messrs L. Pitsi and A. Struyf of the Ceuterick Printing Office.

Particular acknowledgment is due to my Assistant, Drs. F. Van de Ginste, for carefully supervising the preparation of the manuscript and for her skillful and meticulous help in reading proof and compiling the subject index to this volume.

The Editor

TABLE OF CONTENTS

---

* Starred chapters pertain to stochastic systems.

PART II. SPATIAL PROGRAMMING: NETWORKS

PART III. INTERTEMPORAL PROGRAMMING: DYNAMICS

* Starred chapters pertain to stochastic systems.

CHAPTER 0

MATHEMATICAL PROGRAMS FOR ACTIVITY ANALYSIS: INTRODUCTION AND SURVEY

PAUL VAN MOESEKE

Eisdem de rebus volui ad te saepius scribere, non quin confiderem diligentiae tuae, sed rei magnitudo me movebat. Cicero. *Ad Atticum*

The well-worn quip about the camel being a horse designed by a committee expresses a widespread prejudice against collective authorship: "Ainsi voit-on que les bâtiments qu'un seul architecte a entrepris et achevés ont coutume d'être plus beaux et mieux ordonnés que ceux que plusieurs ont tâché de raccommoder". [Descartes. *Discours de la Méthode*.] But while "Mille cerveaux auront beau se coaliser, ils ne composeront jamais le chef-d'œuvre qui sort de la tête d'un Homère", it is admitted in the same passage that "Dans le monde matériel... une multitude arrive plus vite et par différentes routes à la chose qu'elle cherche; des masses d'individus en étudiant chacun de son côté... rencontreront des découvertes dans les sciences, exploreront tous les coins de la création physique". [Chateaubriand. *Mémoires d'Outre-Tombe*].

Where philosopher and literator waver or disagree, the humble programmer may team up with his fellows without apology, especially since there is ample precedent: indeed, the development of mathematical programming, and its application to the analysis of activities, has been marked from the start by a series of symposia reports. (For a list of the earlier ones, see Graves [4, p. 5].)[1] The reason is simple: methods of programming, and of constrained maximization generally, cut a wide swath through a number of disciplines so that economists, management scientists, engineers, systems analysts, and applied mathematicians converge—and confer—on the common techniques. The reader will find both diversity of disciplines and basic unity of approach appropriately reflected in what follows: on the one

[1] *Numbered* references for both chapters 0 and 1 are listed at the end of chapter 1. *Dated* references are found in the Index at the end of the book.

hand, the authors' backgrounds are as varied as the range of topics is broad; on the other hand, all of the papers below are concerned, directly or indirectly, with methods of determining a combination of activities that is optimal on some criterion, or on several criteria, under a number of technical constraints. Thus the volume, in Koopmans' [9, p. 1] words, "testifies to the fundamental unity of the economic problem".

The central production problem, viz. the allocation of scarce resources to desired products, is precisely one of choice of (production) activities under resource constraints. Relating allocation to the core of value theory we show below that, under certain conditions spelled out in chapter 1, an activity choice is efficient if, and only if, there exist nonnegative price systems for, respectively, products and resources, such that efficient choice maximizes revenue. Koopmans [*ibid.*, pp. 1–4] traces the main strands of economic thought that inspired the programming approach to activity analysis, to the discussions of, respectively, the Walras-Cassel equation systems of mathematical economics, the theory of welfare economics, interindustry tables, and transportation problems.

For the reader's convenience we shall now attempt to survey, and to some extent connect, the essays in this collection. The succinct comments below are merely intended as signposts, *not* as summaries, and have been apportioned solely on the basis of the editor's subjective estimate of the need for articulation of separate contributions into a systematic whole.

The three parts of the book follow the logical sequence of *General*, *Spatial*, and *Intertemporal* Programming: they offer techniques designed, respectively, to *indicate*, *locate*, and *date* optimal combinations of activities. Furthermore, within each part chapters have been numbered in approximate order of decreasing generality. Finally, each part includes models of particular interest to the student of stochastic systems: the corresponding chapters (5 and 7 in part I; 9 and 13 in part II; 15, 17, 18, 19, and 20 in part III) have been starred in the Table of Contents.

A mild amount of editorial interference and rearrangement notwithstanding, some overlapping, notational divergence, and terminological variance are, of course, inevitable in a volume of this nature.

1. Mathematical Programming (Part I)

The opening chapter is a theoretical introduction to both the scalar and vectorial activity-analytic models and derives the theory of value dual to efficient allocation. The resulting propositions unify and, we believe, strenghten a number of theorems on constrained maximization (cf. Kuhn and Tucker [10], Uzawa [12], Van Moeseke [15]) and we have been at some pains to provide an adequate background for the duality properties appealed to in all but a few of the subsequent studies.

With the exception of chapter 7 the remaining papers of part I are concerned with advances beyond convex programming.

Chapters 2 and 3 supply algorithms for large classes of general programming problems. Avriel and Zang (ch. 2) group and extend recent results in generalized convexity and its implications for nonlinear programming. The broadening of the customary convex-function concept to the G-convex function consists in using averages other than the weighted arithmetic one. Analogs of theorems on convex functions afford an algorithm for reverse-convex and generalized geometric (signomial) programs. Van Winckel (ch. 3) points out several deficiencies in the classical convergence theorem; he obtains a more general convergence proposition, which he applies to a penalty algorithm: the latter solves the programming problem by transforming it into a sequence of unconstrained problems.

Chapters 4 and 5 focus on the narrower class of geometric programs and are illustrated with numerical examples. Determining the global minimum in signomial programming, Passy (ch. 4) develops a method particularly relevant to engineering design: he delineates the nonconvex constraint set by a union of convex subsets and transforms the original program into a geometric one with conditional constraints. Rijckaert (ch. 5) studies the effects of small parameter perturbations on the optimal values of primal and dual variables: he presents a straightforward scheme, using the quasidual of signomial programs, for adjusting optimal solutions to small perturbations of input data. In fact, the technique applies to both posynomial and signomial problems.

Cutting procedures are used in chapters 6 and 7 in the construction of finite algorithms for integer and homogeneous programming, respectively. Wolsey (ch. 6) derives from a given integer programming problem a wide range of group problems, allowing him to pick within that range a member of arbitrary size, so as to obtain enumerative bounds and generate a variety of valid cuts. Drèze and Van Moeseke (ch. 7) solve the $E-V$ portfolio problem by the truncated-minimax criterion (Van Moeseke, 1965): a finite number of cuts, with a concomitant iteration of parametric quadratic programs, defines a finite algorithm. The approach is, of course, applicable to stochastic linear programming.

2. Spatial Programming: Networks (Part II)

The contributions of part II belong to a family of related problems that can broadly be categorized as network models, a class of programs that generated some of the earliest work (cf. Hitchcock [6], Koopmans [8]) in Operations Research. Transportation models especially have led to the use of network and digraph concepts as tools of analysis.

The first three chapters (8, 9, 10) of this part are general and theoretical in scope.

The comprehensive study by de Ghellinck (ch. 8) provides a systematic treatment of planar networks, emphasizing the symmetry between flows and tensions on such networks, and particularly the possibility of representing any flow as a difference of "face numbers". This property, exploited in conjunction with Dijkstra's (1959) algorithm, leads to the so-called *D* algorithm, here introduced, which is used in turn to generate a class of finite, nondegenerate, nonbacktracking algorithms for, *inter alia*, the max-flow, the transshipment, the min-cost max-flow, and the general piecewise-linear convex problems.

Randomness of arc capacities in transportation networks is dealt with by Doulliez and Jamoulle (ch. 9): capacities are assumed to be independent discrete random variables and flow requirements at the sink nodes are known values. The authors adduce an efficient method, based on a decomposition principle, for finding the probability that all flow requirements be satisfied, as well as the probability that a given arc be found in a minimal cut.

Dragan (ch. 10) studies the general min-cost multicommodity-flow model. In the single-commodity case the necessary and sufficient optimality condition, based on the nonexistence of negative cycles with respect to the flow, is well-known. The optimality condition for the general case is here derived in terms of all feasible systems of cycles. The possibility of basing an efficient algorithm on that condition remains, however, an open question.

The tenor of the remaining three papers of part II is more specific and concrete.

Chapters 11 and 12 deal with what are essentially assignment tasks, respectively of central facilities to population centers, and of schoolrooms to faculty and students. Hansen and Kaufman (ch. 11) construct an implicit-enumeration algorithm for what is termed the "constrained facility-location problem", viz. minimizing transportation costs for the users of a number of facilities, to be established within a given budget at a certain number of feasible locations. In a more expository vein Roubens (ch. 12) surveys existing algorithms for setting up school timetables and proposes two heuristics—based respectively on the concept of float protection and on the Hungarian method—for tackling the problem.

The final chapter of part II by Rubinovitch (ch. 13) explores a probabilistic model for a class of buffers (storage facilities) in data communication networks. Assuming idle and active periods on the input lines to be exponentially distributed, the author calculates the distribution of active periods on the (single) output line as well as the limiting distribution of the data amount in storage.

3. Intertemporal Programming: Dynamics (Part III)

The chapters of part III cover various aspects of, and suggest various paradigms for, sequential programming. All belong to what the economist terms the one-commodity type: essentially, attention shifts from one aspect of activity analysis to another in that, broadly speaking, activity indices, rather than labeling different commodities, now date quantities of the single commodity.

The three opening chapters (14, 15, 16) are theoretical models of dynamic programming.

Richard's (ch. 14) extensive paper deals with the firm facing known requirements over T periods and having to decide, in each, on the amounts to be produced and/or carried over at minimal cost. He generalizes the Hohn-Modigliani (1955) algorithm in several directions by allowing for: (a) variable costs over time; (b) inventory disposal cost; (c) demand functions fluctuating over time. The main analytical tool is standard (primal and dual) convex programming (cf. ch. 1) since nonnegativity constraints on the state variables preclude use of Halkin's (1966) discrete maximum principle.

Dar, Razin, and Yahav (ch. 15) treat a macroeconomic growth model by stochastic dynamic programming. The authors elaborate on the model in Levhari and Srinivasan (1968), viz. maximization of the planner's discounted utility stream in terms of the part c_t of the single commodity k_t consumed in each period t. They consider in particular: the limiting behavior of k_t under certain assumptions; optimal consumption under uncertain lifetimes (random death); and aggregate consumption when initial wealth and discount factors vary among individuals.

The above models of ch. 14 and 15 differ, of course, in that the latter, apart from the introduction of a random element, considers an infinite horizon and is closed (no outside requirements).

The engineering study by Loute (ch. 16), who formulates the minimum-weight design of a turbine disk as a discrete optimal-control problem, belongs to part III from a formal, rather than a substantive point of view and its inclusion illustrates the purview of the nonlinear programming techniques under discussion.

The next two chapters (17, 18) treat sequential planning problems as stochastic systems, viz. queuing and Markov decision models, respectively.

Marchand (ch. 17) considers a facility (e.g. a computer) for whose services customers are queuing up: the latter have no priority claims and are perfectly informed of current congestion and current service price. Under plausible assumptions the author vindicates, and circumscribes, the commonsense notion that Pareto-optimal pricing is monotonic in the degree of congestion. (Pareto-optimal pricing is specified in ch. 1 as an instance of efficient pricing in activity analysis.)

Teugels (ch. 18) takes a different tack in dealing with an important class of consecutive trials as divergent as assembly lines, qualifying examinations, mail delivery chains etc., which share the following features: successive trials are mutually independent but the entire chain of trials has to be repeated as soon as one link fails. As each trial has its own chance of failure, tie, or success, serial tasks of this category reduce to Markov success chains: some classification properties and basic characteristics of such chains are proved.

The closing chapters (19, 20) of part III are rather expository: they conclude the book, fittingly, with a question mark. In communicating their consulting experiences with large organizations, Federwisch (ch. 19) and Wild (ch. 20) trace, in numerical detail, decision sequences arising in the concrete contexts of, respectively, plant capacity planning for a new line of production and hospital laboratory costing for bacteriology tests. In view of the inadequacy of currently available abstract models to deal with intricate systems of this type, further complicated by stochastic features, the authors tentatively resort to a pragmatic approach, applying decision trees and flow charts, respectively.

References

See footnote 1.

PART I

MATHEMATICAL PROGRAMMING

An honest man has hardly need to count more than his ten fingers, or in extreme cases he may add his ten toes, and lump the rest. Simplicity, simplicity, simplicity! I say, let your affairs be as two or three, and not a hundred or a thousand; instead of a million count half a dozen, and keep your accounts on your thumbnail. In the midst of this chopping sea of civilized life, such are the clouds and storms and quicksands and thousand-and-one items to be allowed for, that a man has to live, if he would not founder and go to the bottom and not make his port at all, by dead reckoning, and he must be a great calculator indeed who succeeds. Simplify, simplify!

Henry David THOREAU. Walden

CHAPTER 1

CONSTRAINED MAXIMIZATION AND EFFICIENT ALLOCATION

PAUL VAN MOESEKE

Summary

The article summarizes and strengthens the main saddlepoint results of mathematical programming for both the general and the convex classes of problems. In the typical activity-analytic model the saddlepoint adjoins to the optimal production program certain nonnegative dual variables interpreted as resource (or input) prices. Analogously, the nonnegative weights corresponding to any efficient combination of activities are interpreted as product (or output) prices. Efficient allocation connects both sets of prices.

1. Introduction

This paper is intended as an introduction, both to saddlepoint (or duality) properties in mathematical programming and to their meaning for efficient allocation. Duality properties are referred to in nearly every one of the theoretical papers in this collection. Efficient allocation of resources will be discussed within the general activity-analytic model of production; the latter constitutes a common frame of reference to which the student may relate most of the subsequent discussions in that it links optimality criteria of production back to (dual) resource pricing.

As the present volume by its very scope aims at a wide spectrum of readers this chapter has been written on the expository level, except for the proofs grouped at the end, which assume some familiarity with separation methods and can be passed up by the reader interested primarily in results.

We shall consider the *mathematical-programming problem*,

(P) $$\max f(x) \text{ on } \mathscr{X} = \{x \in X \mid g(x) \geqq 0\},$$

where $f: X \to R$, $g: X \to R^m$. We further define the *Lagrangean* $L: X \times V \to R$ *associated* with P as

$$L(x, v) = f(x) + vg(x),$$

where V is the space of nonnegative m-tuples. (To be quite precise: V is the nonnegative orthant of the conjugate to, or the space of linear functionals on, R^m.) Throughout the chapter scalar products like $vg(x) = \sum v_i g_i(x)$ are denoted by simple juxtaposition (rather than transposition).

One calls (x^*, v^*) a *saddlepoint of L on $X \times V$* if x^*, v^* satisfy

$$L(x, v^*) \underset{X}{\leqq} L(x^*, v^*) \underset{V}{\leqq} L(x^*, v),$$

i.e. if

$$f(x) + v^* g(x) \underset{X}{\leqq} f(x^*) + v^* g(x^*) \underset{V}{\leqq} f(x^*) + vg(x^*). \tag{1.1}$$

The theorems of section 3 will also cover the case where, in addition to the inequality constraint $g(x) \geqq 0$, there is an equality constraint $r(x) = 0$.

2. The Activity-Analytic Model

In what follows we shall distinguish between *two models of activity analysis*, to wit: the *scalar* model, with a single output (scalar maximization); and the *vectorial* model with several outputs (vector maximization), which will be presented as a generalization of the scalar version.

2.1. The Scalar Model

We denote by $f(x)$ the amount of the single desired *output* (revenue, say); by X, the *decision space*; and by $\mathscr{X}$, the set of *feasible* decisions. In the usual case where $X \subset R^n$, a decision x is an n-tuple (*bundle*) of *activities* x_i. Without loss of generality, putting

$$g(x) = b - h(x), \quad b \in R^m, \tag{2.1}$$

reduces the constraint $g(x) \geqq 0$ to

$$h(x) \leqq b, \tag{2.2}$$

where, in the present model, b denotes the m-tuple of *available resource amounts*; $h(x)$, the m-tuple of resource *uses*; and $g(x)$, the m-tuple of *unused supplies* (resource amounts left over at the end of the accounting period to which x refers).

In economic parlance $\mathscr{X}$ can be interpreted as the "technology" or "opportunity set" and its boundary as the "transformation curve". Finally, $v \in V$ is the m-tuple of *dual variables* (or Lagrange multipliers), construed as resource prices[1] *in terms of* the units wherein $f(x)$ is expressed. Consequently, given v^*, the Lagrangean $L(x, v^*) = f(x) + v^* g(x)$ measures the *net worth* of the firm for any decision x, viz. $f(x)$ plus the value of left-over stocks $g(x)$.

Under the conditions

(C) X convex; f, g concave; $g(x^0) > 0$ for some $x^0 \in X$, listed among the premises of some of the propositions in section 3, x^* solves P if, and only if, there is an m-tuple v^* of resource prices satisfying (1.1). *This is a key result which we may paraphrase as follows*:

(a) If x^* is optimal (maximizes revenue) on the set $\mathscr{X}$ of *feasible* decisions then there exists a nonnegative price system v^* such that, by the first inequality in (1.1), x^* *also* maximizes net worth on the *entire* decision space—and not just on the feasible set. The latter point deserves particular emphasis: $x \in X \backslash \mathscr{X}$ (nonfeasibility) means $g_i(x) < 0$, i.e. $h_i(x) > b_i$, for some i, so that $-v_i^* g_i(x) = -v_i^* (b_i - h_i(x))$ is the expenditure, at price v_i^*, for the extra amount of the ith resource. In the case of a resource market with prices v^* one infers that x^* is optimal relatively to *both* production and market operations: put another way, x^* remains optimal even if open-market operations at prices v^* are allowed.

(b) From the second inequality in (1.1) one infers (cf. proof of theorem 3.1 below) that $v^* g(x^*) = 0$ so that $g_i(x^*) > 0$, i.e. $h_i(x^*) < b_i$, implies $v_i^* = 0$ and prices for resources in excess supply vanish. The second inequality in (1.1) thus tells us that v^* is, among all nonnegative price systems, one that *minimizes* the value of supplies. One might interpret this to mean that, while v^* is such that net worth $f(x^*) + v^* g(x^*)$ is maximal on X, this has been accomplished without "window dressing" in that surplus stock has been valued minimally, and indeed at zero worth.

(c) Theorem 3.5 in the last section states that $L(x, v)$ has a saddlepoint if, and only if, it is both a minimax and a maximin:

$$L(x^*, v^*) = \min_V \max_X L(x, v) = \max_X \min_V L(x, v),$$

meaning, by definition, that (x^*, v^*) is a solution to the two-person game with strategy spaces X, V and kernel L. This observation links activity analysis to game theory: maximizing desired output (or revenue) on the

[1] Some alternative terms, such as "internal", or "shadow" prices, suggesting merely accounting, as opposed to market, value convey an institutional limitation inapposite to the present context.

feasible space $\mathscr{X}$ is equivalent to solving concomitantly the dual problems of maximizing net worth on decision space X *and* minimizing the value of excess supplies on the space V of nonnegative resource prices.

(d) It is shown in section 3 that, under conditions C in conjunction with differentiability of f, g, (1.1) is coextensive with the Kuhn-Tucker [10] conditions:

$$(KT)\quad \begin{aligned} &1)\ L_{x^*} = f_{x^*} + v^* g_{x^*} \leqq 0, \\ &2)\ L_{x^*} x^* = (f_{x^*} + v^* g_{x^*}) x^* = 0; \\ &3)\ L_{v^*} = g(v^*) \geqq 0, \\ &4)\ v^* L_{v^*} = v^* g(x^*) = 0. \end{aligned}$$

While KT 3, 4 have already been commented on, KT 1, 2 mean, by (2.1),

$$f_{x^*} - v^* h_{x^*} \leqq 0, \tag{2.3}$$

$$f_{x^*} x^* = v^* h_{x^*} x^*, \tag{2.4}$$

so that

$$x_j^* > 0 \text{ implies } f_{x_j^*} = v^* h_{x_j^*}. \tag{2.5}$$

Interpreting the gradient f_{x^*} as the n-tuple of gross marginal products and $v^* h_{x^*}$ as the n-tuple of marginal costs (both measured in units of the desired output), one identifies the left side in (2.3) as the n-tuple of *net* marginal products so that, by the same formula, no activity yields a positive net marginal product and by (2.5) any activity operated at a positive level breaks even. These conclusions are generalizations of the well-known analogs in *linear* activity analysis (cf. [9, p. 65]).

(e) Neither the "primal solution" x^* nor the "dual[2] solution" v^* to P need be unique, i.e. there may be several saddlepoints. Be (x^*, v^*) and (x^0, v^0) two saddlepoints of L: we show at the end of the chapter that *any dual may be associated with any primal*, i.e. (x^*, v^0) and (x^0, v^*) are also saddlepoints, and further that *L is constant on the set of saddlepoints.* Consequently, net worth $L(x^*, v^*)$ is independent of the particular solution x^* and price system v^* adopted.

---

[2] For dual programming problems with solution v^* see Dorn [1] for quadratic programming; Eisenberg [3] and Van Moeseke [13] for homogeneous programming; Dorn [2], Hanson [5], Wolfe [19], Huard [7], and Van Moeseke [14] for general convex programming. For dual pricing in separable programming see Van Moeseke and de Ghellinck [16].

Saddlepoint and duality properties, as already observed, are protatic to the theoretical, or instrumental in the technical, parts of nearly every one of the papers that follow: the interpretation of the v_i^* as resource prices should, of course, be understood *sensu largo*; in the algorithmic papers, for instance, v_i^* indicates, essentially, the marginal gain from relaxing the ith constraint.

2.2. *The Vectorial Model*

The vectorial activity-analytic model extends P by allowing for *several*, say k, desired outputs: $f(x)$ gets replaced by $F(x)$, where $F: X \to R^k$, and simple maximization by "vector maximization". This is a complete joint-product model. Note further that, for our terminology of *inputs* (b), *decisions* or *activity bundles* (x), and *outputs* $(y = F(x))$, one may substitute *primary*, *intermediate*, and *final commodity bundles*, respectively, thereby drawing attention to the stages of production.

We indicate vector inequalities in the usual way: let $y, z \in R^k$; then $y \geq z$ means $y \geqq z$ *and* $z \neq y$. Clearly, $\geqq$ connotes a partial order on R^k. One calls $F(x^*)$ *vector maximal*, or *efficient*, on $\mathscr{X}$ if $F(x) \geq F(x^*)$ for no $x \in \mathscr{X}$ (i.e. if $F(x^*)$ is maximal relatively to $\geqq$ on the set $Y = \{y \in R^k \mid y = F(x),\ x \in \mathscr{X}\}$.

The foregoing duality and saddlepoint results are directly extended to the vectorial model by *scalarizing*, i.e. maximizing a scalar function: one now solves the mathematical programming problem,

$$\text{(Q)} \qquad \max pF(x) \quad \text{on} \quad \mathscr{X},$$

where p is a semipositive k-tuple, to be interpreted as the k-tuple of *output prices*. Observe that any $p > 0$ leads to an efficient x^*. (*Proof.* For let $F(x^0) \geq F(x^*)$ for some $x^0 \in \mathscr{X}$; then $p > 0$ implies $pF(x^0) > pF(x^*)$, contrary to the assumption that x^* solves Q.)

A more comprehensive proposition on the relationship between the set of nonnegative k-tuples p and the subset of efficient decisions x^* is proved in [15] and merely quoted here; it is rendered intuitively plausible by the figure, which further illustrates the proposition that, if $\mathscr{X}$ is convex (g concave) and F is concave then

$$\text{(2.6)} \qquad Y = \{y \in R^k \mid y \leqq F(x),\ x \in \mathscr{X}\}$$

is convex. (*Proof.* For let $y^1, y^2 \in Y$, i.e. $y^1 \leqq F(x^1)$, $y^2 \leqq F(x^2)$ for some $x^1, x^2 \in \mathscr{X}$. Then, for all $t \in (0, 1)$, $ty^1 + (1-t)y^2 \in Y$ since $ty^1 + (1-t)y^2 \leqq tF(x^1) + (1-t)F(x^2) \leqq F(tx^1 + (1-t)x^2)$, where the latter inequality holds by concavity of F and $tx^1+(1-t)x^2 \in \mathscr{X}$ by convexity of $\mathscr{X}$.) Hence, the image $y^* = F(x^*)$ of every efficient decision x^* can be "reached" by some $p \geq 0$ (i.e. every efficient x^* maximizes $pF(x)$ on $\mathscr{X}$ for some $p \geq 0$). One will, of course, normalize by restricting $p \geq 0$ to the unit simplex.

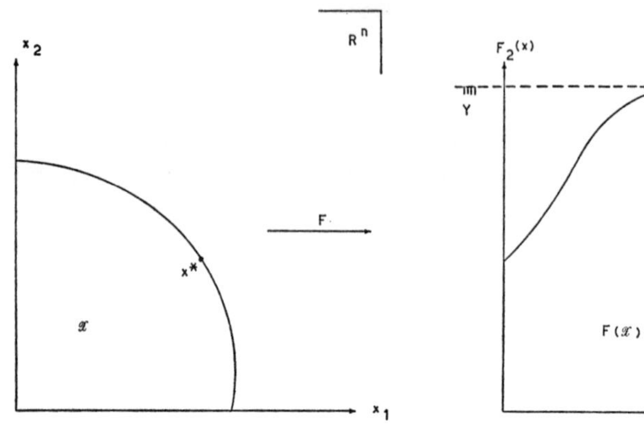

Fig. 1

Proposition 2.1. Let $F: \mathscr{X} \to R^k$. Then

(a) for $p > 0$ (resp. $p \geq 0$) any (resp. some) maximizer x^* of pF on $\mathscr{X}$ is efficient;

(b) of $\mathscr{X}$ is convex and F concave then for any efficient x^* there is a $p \geq 0$ such that x^* maximizes pF on $\mathscr{X}$.

The determination of an efficient x^* has now been reduced to solving the programming problem Q for corresponding $p \geq 0$, with Lagrangean $pF(x) + vg(x)$. Hence, under conditions C, x^* is efficient if, and only if, there is an m-tuple v^* of resource prices satisfying

$$(2.7) \qquad pF(x)+v^*g(x) \underset{X}{\leqq} pF(x^*)+v^*g(x^*) \underset{V}{\leqq} pF(x^*)+vg(x^*).$$

By proposition 2.1 a k-tuple p of *output prices* can be associated with any efficient decision and vice versa: the m-tuple of *input prices* v^* is then determined by (2.7).

The general activity-analytic model is, of course, amenable to a wide range of interpretations: some of these, including the statistical theory of Bayesian and admissible decision rules, have been set forth in some detail in [15, 17]. We mention here that the efficiency concept is central to Welfare Economics, where $\mathscr{X}$ represents the technology; F, the k-tuple of individual utility functions; and where an efficient decision is a *Pareto optimum*: the activity-analytic model is actually given this interpretation by Marchand (ch. 17) and by Dar, Razin, and Yahav (ch. 15); in both chapters a Pareto optimum is arrived at by maximizing a weighted average pF of individual utilities F_i. Again, in risk programming (cf. Drèze and Van Moeseke (ch. 7) and Federwisch (ch. 19)) $\mathscr{X}$ is the budget set and efficient decisions are here called Markowitz [11] efficient; here F has two coordinates: one identifies F_1 as the expectation, and $(-F_2)$ as the standard deviation, of returns.

3. Saddlepoint Theorems

This section provides proofs for the saddlepoint properties referred to above. In fact, we extend the programming model to the case where, in addition to the inequality constraints $g(x) \geqq 0$, there are *equality* constraints $r(x) = 0$, where $r: X \to R^q$. The first four propositions relate to general, the next four to convex programming, and the final ones to abstract saddlepoint properties appealed to in paragraphs (c) and (e) of the foregoing section.

Taking the equality constraints into account will require consideration of the programming problem P' and the associated expressions L', SP', KT', C', D' parallel to P and L, SP, KT, C, D. Both programming problems have been tabulated here for ready reference (table 1).

In what follows the variables x, v, u belong to X, V, and U, respectively. Proofs will be supplied for the following propositions: theorems 3.1, 3.2 etc. apply to P; theorems 3.1′, 3.2′ etc. are the corresponding analogs for P'.

GENERAL PROGRAMMING

THEOREM 3.1. If x^*, v^* satisfy SP then x^* solves P.

THEOREM 3.2. If x^*, v^* satisfy SP, D then x^*, v^* satisfy KT.

THEOREM 3.1′. If x^*, v^*, u^* satisfy SP', then x^* solves P'.

THEOREM 3.2′. If x^*, v^*, u^* satisfy SP', D' then x^*, v^*, u^* satisfy KT'.

TABLE 1

PROBLEM P

(P)	$\max f(x)$ on $\mathscr{X}=\{x\in X \mid g(x)\geqq 0\}$
(L)	$L(x,v)=f(x)+vg(x)$
(SP)	$f(x)+v^*g(x)\underset{X}{\overset{(1)}{\leqq}} f(x^*)+v^*g(x^*)\underset{V}{\overset{(2)}{\leqq}} f(x^*)+vg(x^*)$
(KT)	1) $L_{x^*}\leqq 0$ 2) $L_{x^*}x^*=0$ 3) $L_{v^*}\geqq 0$ 4) $v^*L_{v^*}=0$
(C)	X convex; f, g concave; $g(x^0)>0$, some $x^0\in X$
(D)	$X=R^n_{\neq}$; f, g differentiable

PROBLEM P'

(P')	$\max f(x)$ on $\mathscr{X}'=\{x\in X \mid g(x)\geqq 0,\ r(x)=0\}$
(L')	$L'(x,v,u)=f(x)+vg(x)+ur(x)$
(SP')	$f(x)+v^*g(x)+u^*r(x)\underset{X}{\overset{(1)}{\leqq}} f(x^*)+v^*g(x^*)+u^*r(x^*)\underset{V\times U}{\overset{(2)}{\leqq}} f(x^*)+vg(x^*)+ur(x^*)$
(KT')	1) $L_{x^*}\leqq 0$ 2) $L_{x^*}x^*=0$ 3) $L_{v^*}\geqq 0$ 4) $v^*L_{v^*}=0$ 5) $L_{u^*}=0$
(C')	X convex; f, g concave; r linear; $g(x^0)>0$ and $r(x^0)=0$, some $x^0\in X$
(D')	$X=R^n_{\neq}$; f, g, r differentiable

In this table f: $X\to R$, g: $X\to R^m$, r: $X\to R^q$, L: $X\times(V\times U)\to R$; V is the set of nonnegative real m-tuples; U is the set of real q-tuples.

CONVEX PROGRAMMING

THEOREM 3.3. Let C hold.[3] For any solution x^* of P there is a v^* such that x^*, v^* satisfy SP.

THEOREM 3.4. Let C, D hold. If x^*, v^* satisfy KT they satisfy SP.

THEOREM 3.3′. Let C' hold. For any solution x^* of P there are v^*, u^* such that x^*, v^*, u^* satisfy SP'.

THEOREM 3.4′. Let C', D' hold. If x^*, v^*, u^* satisfy KT' they satisfy SP'.

COROLLARY 3.5. Let C hold. Then x^* solves P iff x^* satisfies SP for some v^*. (Follows from theorems 3.1 and 3.3.)

COROLLARY 3.6. Let C, D hold. Then x^*, v^* satisfy KT iff they satisfy SP. (Follows from theorems 3.2 and 3.4.)

COROLLARY 3.7. Let C, D hold. Then x^* solves P iff there is a v^* such that x^*, v^* satisfy KT. (Follows from corollaries 3.5 and 3.6.)

COROLLARY 3.5′. Let C' hold. Then x^* solves P' iff x^* satisfies SP' for some v^*, u^*. (Follows from theorems 3.1′ and 3.3′.)

COROLLARY 3.6′. Let C', D' hold. Then x^*, v^*, u^* satisfy KT' iff they satisfy SP'. (Follows from theorems 3.2′ and 3.4′.)

COROLLARY 3.7′. Let C', D' hold. Then x^* solves P' iff there are v^*, u^* such that x^*, v^*, u^* satisfy KT'. (Follows from corollaries 3.5′ and 3.6′.)

GENERAL SADDLEPOINT AND MINIMAX PROPERTIES

Be X, V any two spaces and L any real-valued function on $X \times V$, denote by $S \subset X \times V$ the set of saddlepoints of L on $X \times V$, and assume that both min max $L(x, v)$ and max min $L(x, v)$ exist. We shall prove:

LEMMA 3.a. $\min_V \max_X L(x, v) \geqq \max_X \min_V L(x, v)$.

THEOREM 3.b. S is nonempty iff $\min_V \max_X L(x, v) = \max_X \min_V L(x, v)$.

THEOREM 3.c. L is constant on S.

THEOREM 3.d. S is a *rectangle* in $X \times V$, i.e., if S owns (x^*, v^*) and (x^0, v^0) then S also owns (x^*, v^0) and (x^0, v^*).

---

[3] C may be relaxed in the case of homogeneous programming, where the condition $g(x^0) > 0$ can be dropped, cf. Van Moeseke [18].

PROOFS

Proof of theorem 3.1.

(a) $v^*g(x^*)=0$; for putting $v=0$ in $SP(2)$ reveals that $v^*g(x^*) \leqq 0$. Also, $v^*g(x^*) \geqq 0$ because $g(x^*) \geqq 0$: indeed, if $g_i(x^*)<0$, some i, then taking $v_i>0$ large enough would make $v_i g_i(x^*)$, hence $vg(x^*)$, negative and large enough in absolute value to contradict $SP(2)$.

(b) x^* solves P: as $v^*g(x^*) \geqq 0$, all $x \in \mathscr{X}$, $SP(1)$ by (a) implies $f(x) \leqq f(x^*)$, all $x \in \mathscr{X}$.

Proof of theorem 3.2. We prove only KT 1, 2. (The proof of KT 3,4 is analogous.)

1) $L_{x^*} \leqq 0$. For let $L_{x_i^*}>0$, some i; then $L(\cdot, v^*)$ may be increased by increasing x_i in a neighborhood of x_i^*, contrary to $SP(1)$.

2) $L_{x^*}x^*=0$: for if $\neq 0$ then, for at least one i, $L_{x_i^*}x_i^* \neq 0$, i.e. $L_{x_i^*}<0$ and $x_i^*>0$ so that $L(\cdot, v^*)$ may be increased by decreasing x_i in a neighborhood of x_i^*, contrary to $SP(1)$.

Proof of theorem 3.1'.

(a) $v^*g(x^*)+u^*r(x^*)=0$. For $r(x^*)=0$ because $r_i(x^*)>0$ (resp. <0) and $u_i<0$ (resp. >0) and large enough in absolute value makes $u_i r_i(x^*)<0$ and large enough in absolute value to contradict $SP'(2)$. As $r(x^*)$ vanishes from both sides of $SP'(2)$ one merely has to show that $v^*g(x^*)=0$ (same argument as in the proof of theorem 3.1 sub (a)).

(b) x^* solves P': indeed, for all $x \in \mathscr{X}$, $r(x)=0$ so that, for all $x \in \mathscr{X}$, $SP'(1)$ reduces to $SP(1)$ and the argument is the same as in the proof of theorem 3.1 sub (b).

Proof of theorem 3.2'. The proof of KT' 1, 2, 3, 4 is like in theorem 3.2. We prove only KT' 5, viz. $L_{u^*}=0$. Indeed if $L_{u_i^*}>0$ (resp. <0), some i, then $L(x^*, v^*, \cdot)$ can be decreased by decreasing (resp. increasing) u_i in a neighborhood of u_i^*, contrary to $SP'(2)$.

Proof of theorem 3.3. The sets[4]

$$A=\{(z_0, z) \in R^{1+m} \mid (z_0, z) \leqq (f(x), g(x)), \text{ some } x \in X\},$$
$$B=\{(z_0, z) \in R^{1+m} \mid (z_0, z) > (f(x^*), 0_m)\},$$

[4] The origin of, say, R^m is henceforth denoted 0_m if there is any danger of confusion; otherwise 0 suffices.

are convex (A by application of the proof following (2.6); B is a displaced nonnegative orthant), do not intersect, and have a common boundary point $(f(x^*), 0_m)$. Hence, there is[5] through that point a separating hyperplane with normal (y_0, y): by choosing B above the hyperplane necessarily

$$(y_0, y) \geq 0 \tag{3.1}$$

since B is a displaced nonnegative orthant. One has

$$y_0 f(x) + yg(x) \underset{X}{\leqq} y_0 f(x^*) + 0. \tag{3.2}$$

Now $y_0 > 0$ for, if $y_0 = 0$ then $y \geq 0$ by (3.1) so $yg(x) \underset{X}{\leqq} 0$; but, by C, $yg(x^0) > 0$, a contradiction.

Dividing (3.2) through by $y^0 > 0$ and putting $v^* = y/y^0$ yields

$$f(x) + v^* g(x) \underset{X}{\leqq} f(x^*).$$

For $x = x^*$ this reduces to $v^* g(x^*) \leqq 0$. As $x^* \in \mathscr{X}$ one has $g(x^*) \geqq 0$, hence also $v^* g(x^*) \geqq 0$, so $v^* g(x^*) = 0$. As $v \geqq 0$ on V one obtains also $SP(2)$.

Proof of theorem 3.4. As $L(x^*, v^*)$ is concave in x one has

$$L(x, v^*) - L(x^*, v^*) \underset{X}{\leqq} L_{x^*}(x - x^*). \tag{3.3}$$

Further $L_{x^*} x \leqq 0$ and $L_{x^*} x^* = 0$ by $KT\,1$ and 2, respectively, so that $L_{x^*}(x - x^*) \leqq 0$ and, by (3.3),

$$L(x, v^*) \underset{X}{\leqq} L(x^*, v^*).$$

Analogously, the second saddlepoint inequality $L(x^*, v^*) \underset{V}{\leqq} L(x^*, v)$ follows from $KT\,3$, 4 using the fact that $L(x^*, v)$ is linear, hence convex, in v.

Proof of theorem 3.3′. The sets

$$A' = \{(z_0, z, z') \in R^{1+m+q} \mid (z_0, z) \leqq (f(x), g(x)),\ z' = r(x), \text{ some } x \in \mathscr{X}\},$$
$$B' = \{(z_0, z, z') \in R^{1+m+q} \mid (z_0, z) > (f(x^*), 0_m),\ z' = 0_q\},$$

are convex (cf. proof of theorem 3.3), do not intersect, and have a common boundary point $(f(x^*), 0_m, 0_q)$. Hence, there is through that point a separat-

[5] This separation technique was introduced by Uzawa [12].

ing hyperplane H with normal (y_0, y, y'): by choosing B' above H one has

$$(3.4) \qquad y_0 f(x)+yg(x)+y'r(x) \underset{X}{\leqq} y_0 f(x^*)+0+0.$$

Denote $Q=\{(z_0, z, z')\in R^{1+m+q} \mid z'=0_q\}$, call $A''=A'\cap Q$, $B''=B'\cap Q$, and observe that the plane $H\cap Q$ separates A'', B''. As B'' is a translated nonnegative orthant in R^{1+m} necessarily

$$(3.5) \qquad (y_0, y)\geq 0.$$

Now $y_0>0$ for, if $y_0=0$ then $y\geq 0$ and by (3.4) $yg(x^0)+y'r(x^0)\leqq 0$; but by C', $r(x^0)=0$ so that $yg(x^0)\leqq 0$, contrary to $g(x^0)>0$.

Dividing (3.4) through by $y_0>0$ and putting $v^*=y/y^0$, $u^*=y'/y_0$ yields

$$f(x)+v^*g(x)+u^*r(x) \underset{X}{\leqq} f(x^*).$$

For $x=x^*$ this reduces to $v^*g(x^*)+u^*r(x^*)\leqq 0$. As $x^*\in\mathscr{X}'$ one has $g(x^*)\geqq 0$, $r(x^*)=0$ so that $v^*g(x^*)+u^*r(x^*)=0$. As further $v\geqq 0$ on V one obtains also $SP'(2)$.

Proof of theorem 3.4′.

(a) As f, g, r are concave in x, so is L; hence the reasoning of the proof of theorem 3.4 applies to the effect that

$$L(x, v^*, u^*) \underset{X}{\leqq} L(x^*, v^*, u^*).$$

(b) By linearity, hence convexity, of L in v, u one has

$$(3.6) \qquad L(x^*, v, u)-L(x^*, v^*, u^*)\geqq((v, u)-(v^*, u^*))L_{(v^*, u^*)}$$
$$=vL_{v^*}+uL_{u^*}-v^*L_{v^*}-u^*L_{u^*}.$$

By KT' 3, 4, 5 and $v\geqq 0$ one has $vL_{v^*}\geqq 0$, while the three last terms on the right in (3.6) vanish so that the second saddlepoint inequality follows:

$$L(x^*, v^*, u^*) \underset{V\times U}{\leqq} L(x^*, v, u).$$

In the next two propositions the operators max, min have domains X and V, respectively.

Proof of lemma 3.a. Let

$$(3.7) \qquad \begin{aligned} L(\bar{x}, \bar{v}) &= \min\max L(x, v),\\ L(\hat{x}, \hat{v}) &= \max\min L(x, v).\end{aligned}$$

By definition

$$(3.8) \qquad L(\bar{x}, \bar{v}) \geqq L(\hat{x}, \bar{v}) \quad \text{and} \quad L(\hat{x}, \hat{v}) \leqq L(\hat{x}, \bar{v}).$$

Combining both inequalities yields $L(\hat{x}, \hat{v}) \leqq L(\bar{x}, \bar{v})$ and the desired result follows.

Proof of theorem 3.b. In view of lemma 3.a we merely establish that S is nonempty iff

$$(3.9) \qquad \min \max L(x, v) \leqq \max \min L(x, v).$$

Necessity. Be $(x^*, v^*) \in S$. Then by definition

$$\min \max L(x, v) \leqq \max L(x, v^*) = L(x^*, v^*) = \min L(x^*, v) \leqq \max \min L(x, v).$$

Sufficiency. Reverting to notation (3.7) one has

$$L(x, \bar{v}) \underset{X}{\overset{(1)}{\leqq}} L(\bar{x}, \bar{v}) \overset{(2)}{\leqq} L(\hat{x}, \hat{v}) \overset{(3)}{\leqq} L(\hat{x}, \bar{v}) \overset{(4)}{\leqq} L(\bar{x}, \bar{v}) \overset{(5)}{\leqq} L(\hat{x}, \hat{v}) \underset{V}{\overset{(6)}{\leqq}} L(\hat{x}, v),$$

where inequalities 1,6 hold by (3.7); 2,5 by (3.9); 3,4 by (3.8). Comparison of the center and terminals of the inequality chain indicates that $(\hat{x}, \bar{v}) \in S$.

Proof of theorem 3.c. Be (x^*, v^*), (x^0, v^0) two arbitrary points in S. Then by repeated application of the definition of a saddlepoint,

$$(3.10) \qquad L(x^*, v^*) \leqq L(x^*, v^0) \leqq L(x^0, v^0) \leqq L(x^0, v^*) \leqq L(x^*, v^*).$$

All terms of the chain have the same value by identity of terminals: in particular, $L(x^*, v^*) = L(x^0, v^0)$.

Proof of theorem 3.d. By repeated application of the definition of a saddlepoint,

$$L(x, v^0) \underset{X}{\leqq} L(x^0, v^0) = L(x^*, v^0) = L(x^*, v^*) \underset{V}{\leqq} L(x^*, v),$$

where the equalities hold because all terms in (3.10) are equal. Hence $(x^*, v^0) \in S$. The proof of $(x^0, v^*) \in S$ is analogous. Consequently, S is rectangular.

References

1. Dorn, W. S. Duality in quadratic programming. *Quarterly of Applied Mathematics*, **18**: 155–62, 1960.

2. DORN, W. S. A duality theorem for convex programs. *IBM Research Journal*, **4**: 407–13, 1960.
3. EISENBERG, E. Duality in homogeneous programming. *Proceedings of the American Mathematical Society*, **12**: 783–87, 1963.
4. GRAVES, R. L. ed. *Recent Advances in Mathematical Programming*. McGraw-Hill, New York, 1963.
5. HANSON, M. A. A duality theorem in nonlinear programming with nonlinear constraints. *Australian Journal of Statistics*, **3**: 64–71, 1961.
6. HITCHCOCK, F. L. The distribution of a product from several sources to numerous localities. *Journal of Mathematics and Physics*, **20**: 224–30, 1941.
7. HUARD, P. Dual programs. *IBM Journal of Research and Development*, **6**: 137–39, 1962.
8. KOOPMANS, T. C. Optimum utilization of the transportation system. *Proceedings of the International Statistical Conferences*, vol. 5. Washington, 1947.
9. KOOPMANS, T. C., ed. *Activity Analysis of Production and Allocation*. Wiley, New York, 1951.
10. KUHN, H. W. and TUCKER, A. W. Nonlinear programming. *Proceedings of the Second Berkeley Symposium on Mathematical Statistics and Probability*. University of California Press, 1951.
11. MARKOWITZ, H. M. *Portfolio Selection*. Wiley, New York, 1959.
12. UZAWA, H. The Kuhn-Tucker theorem in concave programming. ARROW, K. J., HURWICZ, L. and UZAWA, H., eds. *Studies in Linear and Nonlinear Programming*. Stanford University Press, 1958.
13. VAN MOESEKE, P. Stochastic linear programming. *Yale Economic Essays*, **5**: 196-254, 1965.
14. VAN MOESEKE, P. A general duality theorem of convex programming. *Metroeconomica*, **17**: 161–70, 1965.
15. VAN MOESEKE, P. Towards a theorey of efficiency. QUIRK, J. and ZARLEY, A., eds. *Papers in Quantitative Economics*, vol. 1. University Press of Kansas, 1968.
16. VAN MOESEKE, P. and de GHELLINCK, G. Decentralization in separable programming. *Econometrica*, **37**: 73–8, 1969.
17. VAN MOESEKE, P. Efficient decisions in economics and statistics. *Alumni* (Journal of the Belgian Science Foundations), **40**: 19–29, 1970.
18. VAN MOESEKE, P. Saddlepoint in homogeneous programming without Slater condition. *Econometrica*, **42** (forthcoming), 1974.
19. WOLFE, P. A duality theorem for nonlinear programming. *Quarterly of Applied Mathematics*, **19**: 239-44, 1961.

CHAPTER 2

GENERALIZED CONVEX FUNCTIONS WITH APPLICATIONS TO NONLINEAR PROGRAMMING

MORDECAI AVRIEL and ISRAEL ZANG

Summary

Generalizations of the notion of convexity, unifying some previously derived results, are introduced. Analogues of theorems on convex functions are presented for the general case. These theorems are used for deriving an algorithm for solving a certain class of nonlinear programs. The proposed algorithm extends results on existing algorithms for reverse-convex and generalized geometric (signomial) programs.

1. Introduction

Growing interest in nonlinear nonconvex optimization problems has led to extensions of the notion of convexity and to weakening the convexity assumptions which are sufficient but not always necessary for obtaining significant results in theory and applications. The present work unifies and generalizes recent results on generalized convexity and its applications in nonlinear programming [1, 2]. This extension is based on generalizing the definition of ordinary convexity by employing averages other than the weighted arithmetic average, used in convex functions. It is hoped that this extension will also lead to further research in generalizing additional results limited so far to convex functions and programming.

2. Definitions

Let G be a continuous real-valued strictly monotonic function defined on $D \subset R$, where R denotes the real line. Let q_1, q_2 be nonnegative numbers, called weights, satisfying $q_1 + q_2 = 1$. A real-valued function f defined on a convex set $C \subset R^n$ is said to be *G-convex* if for any $x^1 \in C$, $x^2 \in C$, q_1 and q_2 we have

$$f(q_1 x^1 + q_2 x^2) \leq G^{-1}[q_1 Gf(x^1) + q_2 Gf(x^2)], \tag{2.1}$$

where G^{-1} is the inverse of G. In order to make the above inequality meaningful we assume that the range of f is contained in D. Note that the conditions imposed on G assure the existence of G^{-1} which is also continuous and strictly monotonic.

Let us relate now the above inequality to ordinary convexity and to some earlier results on generalized convexity. Suppose that $G(x)=x$, $x \in R$. Then $G^{-1}(x)=x$, $x \in R$. Substituting into (2.1) yields

$$f(q_1 x^1 + q_2 x^2) \le q_1 f(x^1) + q_2 f(x^2) \tag{2.2}$$

and f is a convex function in the ordinary sense.

Suppose now that $G(x)=\mathrm{e}^{rx}$, $x \in R$, where $r \neq 0$ is a real parameter. Then $G^{-1}(x)=\log x^{1/r}$, $x>0$ and from (2.1) we obtain

$$f(q_1 x^1 + q_2 x^2) \le \begin{cases} \log [q_1 \mathrm{e}^{rf(x^1)} + q_2 \mathrm{e}^{rf(x^2)}]^{1/r} & r \neq 0 \\ q_1 f(x^1) + q_2 f(x^2) & r = 0. \end{cases} \tag{2.3}$$

Functions satisfying (2.3) are called r-convex functions. See Avriel [1] and Martos [9].

Take now $G(x)=x^r$, $x>0$ and $r \neq 0$ is again a real number.

Then $G^{-1}(x)=x^{1/r}$, $x>0$ and (2.1) becomes

$$f(q_1 x^1 + q_2 x^2) \le \begin{cases} [q_1 f^r(x^1) + q_2 f^r(x^2)]^{1/r} & r \neq 0 \\ f^{q_1}(x_1) f^{q_2}(x_2) & r = 0. \end{cases} \tag{2.4}$$

Positive functions f satisfying (2.4) were called r^+-convex functions by Avriel [2]. The special case of $r=0$ in (2.4) was also treated by Klinger and Mangasarian [6].

These two examples can be combined by letting $G=F^r$, where F is a positive function such that the assumptions made on G at the beginning of this section are satisfied. Then, f is said to be F_r-convex if for any $x^1 \in C$, $x^2 \in C$ and q we have

$$f(q_1 x^1 + q_2 x^2) \le \begin{cases} F^{-1}\{q_1 F^r[f(x^1)] + q_2 F^r[f(x^2)]\}^{1/r} & r \neq 0 \\ F^{-1}\{F^{q_1}[f(x^1)] F^{q_2}[f(x^2)]\} & r = 0. \end{cases} \tag{2.5}$$

Functions which are *G-concave* are defined by reversing the sense of the inequality in (2.1). We shall study only G-convex functions in this paper since analogous results for G-concave functions can be easily derived from our work.

3. Properties of G-Convex Functions

The righthand side of the inequality in (2.1) is called the G-mean of the function values $f(x^1), f(x^2)$. Properties of general means of this type can be found in Hardy, Littlewood and Polya [5]. It is shown there that for continuous and strictly monotonic functions G there is a unique number M such that

$$G(M) = q_1 \, Gf(x^1) + q_2 \, Gf(x^2) \tag{3.1}$$

and

$$\min[f(x^1), f(x^2)] \leq M \leq \max[f(x^1), f(x^2)] \tag{3.2}$$

where equality holds if and only if $f(x^1) = f(x^2)$.

We can prove now

THEOREM 3.1. Every G-convex function on a convex set C is also quasiconvex on C.

Proof. A function f is quasiconvex on a convex set C if for any $x^1 \in C$, $x^2 \in C$ and q

$$f(q_1 x^1 + q_2 x^2) \leq \max[f(x^1), f(x^2)]. \tag{3.3}$$

If f is G-convex, then by (2.1), (3.1) and (3.2)

$$f(q_1 x^1 + q_2 x^2) \leq G^{-1} G(M) = M \leq \max[f(x^1), f(x^2)]. \tag{3.4}$$

QED.

In [1] it was shown that there is a continuous transformation from the family of convex functions to the family of quasiconvex functions via the concept of r-convex functions. This result can be generalized as will be demonstrated below. First we need a lemma, due to Hardy, Littlewood and Polya [3].

Let $\quad w \in R^n, \; w = (w_1, \ldots, w_n), \; q \in R^n, \; q = (q_1, \ldots, q_n)$

such that

$$q_j \geq 0, \quad q_1 + q_2 + \ldots + q_n = 1. \tag{3.5}$$

Define

$$M_G(w; q) = G^{-1}[q_1 \, G(w_1) + \ldots + q_n \, G(w_n)]. \tag{3.6}$$

Then we have

LEMMA 3.2. Let G_1 and G_2 be strictly monotonic and let $\Gamma = G_2 G_1^{-1}$. Suppose that Γ possesses a second derivative Γ'' such that $\Gamma''(x) > 0$ for all x whenever G_2 is increasing and $\Gamma''(x) < 0$ whenever G_2 is decreasing. Then,

$$M_{G_1}(w; q) \leq M_{G_2}(w; q) \tag{3.7}$$

with equality holding if and only if $w_1 = w_2 = \ldots = w_n$.

From this lemma we get immediately

THEOREM 3.3. Let G_1, G_2 and Γ satisfy the hypotheses in lemma 3.2. Then, every G_1-convex function f is also G_2-convex.

The last result can be slightly weakened. It can be shown that (3.7) holds if and only if $\Gamma = G_2 G_1^{-1}$ is a convex function. We can say, therefore, that if Γ is convex then every G_1-convex function is also G_2-convex. This covers the trivial case of $G_1 = G_2$. Taking $G_1 = \log x$, $G_2 = x$ we get $\Gamma = e^x$. A G_1-convex positive function f satisfies

$$f(q_1 x^1 + q_2 x^2) \leq f^{q_1}(x^1) f^{q_2}(x^2) \tag{3.8}$$

and a G_2-convex function satisfies

$$f(q_1 x^1 + q_2 x^2) \leq q_1 f(x^1) + q_2 f(x^2). \tag{3.9}$$

It follows that if f is G_1-convex (log-convex by [6] and 0^+-convex by [2]) then it is also convex.

Suppose now that we have a sequence $\{G\}$ of continuous strictly increasing functions $\{G\} = G_1, G_2, \ldots$ Let $\Gamma_k = G_{k+1} G_k^{-1}$ such that $\Gamma_k''(x) > 0$ for all x and $k = 1, 2, \ldots$.

Then it follows from lemma 3.2 that

$$\lim_{k \to \infty} M_{G_k}(w_j q) = \max\,[w_1, w_2, \ldots, w_n] \tag{3.10}$$

and by choosing a suitable sequence (such as $\{G\} = e^x, e^{x^2}, e^{x^3}, \ldots$) one can get a gradual transformation from convex functions to quasiconvex functions.

Every G-convex function can be transformed into an ordinary convex or concave function as can be seen in the following result.

THEOREM 3.4. Let f be a real-valued function defined on a convex set $C \subset R^n$ and let G be a real-valued strictly monotonic function defined on a subset D of R which contains the range of f. Let $\hat{f}$ be defined by

(3.11) $$\hat{f}(x) = Gf(x), \quad x \in C.$$

Then f is G-convex on C if and only if $\hat{f}$ is convex whenever G is increasing and concave whenever G is decreasing.

Proof. Follows directly from the definition of G-convex functions. QED.

A function f is convex in the ordinary sense if and only if $g = -f$ is concave. We shall see now that this result can be generalized for G-convex functions. Suppose that G is a one-one map of $S \subset R$ onto $T \subset R$ and can be written as a composite $G = \phi\psi$, where ϕ and ψ are one-one maps of S onto $U \subset R$ and U onto T, respectively. Then $G^{-1} = \psi^{-1}\phi^{-1}$ [7] and we have

THEOREM 3.5. Let $G = \phi\psi$ satisfy the above assumptions and suppose that ψ is a decreasing function. Then f is G-convex if and only if ψf is ϕ-concave.

Proof. Suppose that f is G-convex and $G = \phi\psi$, i.e.

(3.12) $$f(q_1 x^1 + q_2 x^2) \le \psi^{-1}\phi^{-1}[q_1 \phi\psi f(x^1) + q_2 \phi\psi f(x^2)].$$

Since ψ is decreasing

(3.13) $$\psi f(q_1 x^1 + q_2 x^2) \ge \phi^{-1}[q_1 \phi(\psi f(x^1)) + q_2 \phi(\psi f(x^2))]$$

and ψf is ϕ-concave. The proof of the converse statement is similar. QED.

As a special case of the theorem consider first $G(x) = x$ with $\phi(x) = -x$ and $\psi(x) = -x$. Then f is G-convex (i.e. ordinary convex) if and only if $\psi f = -f$ satisfies

(3.14) $$-f(q_1 x^1 + q_2 x^2) \ge -[q_1 f(x^1) + q_2 f(x^2)] = q_1(-f(x^1)) + q_2(-f(x^2)),$$

i.e. $-f$ is concave. Let now $G(x) = e^{rx}$, $r \ne 0$, with $\phi(x) = e^{-rx}$ and $\psi(x) = -x$. Then f is r-convex if and only if $\psi f = -f$ satisfies

(3.15) $$-f(q_1 x^1 + q_2 x^2) \ge \log[q_1 e^{(-r)(-f(x^1))} + q_2 e^{(-r)(-f(x^2))}]^{1/-r},$$

i.e. $-f$ is $(-r)$-concave.

Unfortunately, some elementary algebraic properties of convex functions do not have their analogues for G-convex functions without further restricting G. Thus, if f is G-convex, then $f+\alpha$ or βf where α and β are real and positive constants, respectively, are not necessarily G-convex. More generally, if f and g are G-convex functions, their nonnegative linear combination is not necessarily G-convex.

There are, however, properties of convex functions which do generalize to G-convex functions. For example, it is easy to show that if f_i, $i \in I$ is a family of G-convex functions on $C \subset R^n$ such that G is increasing, then the function f defined by

$$f(x) = \sup_{i \in I} f_i(x) \tag{3.16}$$

is G-convex on C. Similarly, one can easily prove that f is G-convex on $C \subset R^n$ if and only if for every $x^1 \in C$, $x^2 \in C$, the function ϕ given by

$$\phi(\lambda) = f(\lambda x^1 + (1-\lambda)x^2) \tag{3.17}$$

is G-convex for $0 \leq \lambda \leq 1$.

Some important properties of convex functions from an optimization point of view also have their counterpart for G-convex functions. For example, if G is monotone increasing then every local minimum of a G-convex function is a global minimum and the set of global minima is convex.

In the case of differentiable G-convex functions one can derive several generalizations of results for convex functions. We shall only mention here an alternative definition of twice differentiable G-convex functions. Let G be twice differentiable and monotone increasing. Then the function f defined in an open convex set $C \subset R^n$ is G-convex if and only if the matrix Q given by

$$Q(x) = G'(f(x)) \nabla^2 f(x) + G''(f(x)) \nabla f(x) (\nabla f(x))^T \tag{3.18}$$

is positive semidefinite for every $x \in C$.

It is clear that for $G(x) = x$ we get the well-known result for convex functions and for $G(x) = e^{rx}$ we get the alternative definition of r-convex functions [1].

A differentiable real-valued function ϕ is said to be pseudo-convex [8] on a convex set $C \subset R^n$ if for each $x^1 \in C$, $x^2 \in C$

$$(x^2 - x^1)^T \nabla \phi(x^1) \geq 0 \text{ implies } \phi(x^2) \geq \phi(x^1). \tag{3.19}$$

We have then

THEOREM 3.6. Let f be a differentiable G-convex function on a convex set $C \subset R^n$ and let G be differentiable. Then f is pseudoconvex on C.

Proof. Suppose that G is increasing. Then, by Theorem 3.4 the function Gf is convex. Hence, for each $x^1 \in C$, $x^2 \in C$

$$Gf(x^2) \geq Gf(x^1) + (x^2 - x^1)^T G'(f(x^1)) \nabla f(x^1). \tag{3.20}$$

Suppose now that $(x^2-x^1)^T \nabla f(x^1) \geq 0$. Since G is increasing we have $G'(f(x^1)) \geq 0$, hence $(x^2-x^1)^T G'(f(x^1)) \nabla f(x^1) \geq 0$ and $Gf(x^2) \geq Gf(x^1)$. It follows that $f(x^2) \geq f(x^1)$ and f is pseudoconvex. The proof for a decreasing G is similar. QED.

4. Nonlinear Programs with G-Convex Functions

Consider now a nonlinear program P

$$\min \phi(x) \tag{4.1}$$

subject to $x \in C$ and

$$\psi(x) \leq 0, \tag{4.2}$$

where ϕ is a real-valued convex function, $C \subset R^n$ is a convex set and ψ is a vector-valued function, whose components are not necessarily quasiconvex. The feasible set of program P is $S = C \cap X$, where

$$X = \{x : \psi(x) \leq 0\}. \tag{4.3}$$

The set S is generally nonconvex. A solution of P will be any point $x^* \in S$ satisfying the Kuhn-Tucker necessary conditions for a local optimum. We shall present an algorithm for finding a solution of P by solving a sequence of convex programs, provided that ψ belongs to a special class of functions. Let $\psi = (\psi_1, \ldots, \psi_m)$ and suppose that each ψ_i is given by

$$\psi_i(x) = \sum_{j=1}^{J(i)} f_{ij}(x) \qquad i = 1, \ldots, m, \tag{4.4}$$

where the f_{ij} are G_{ij}-convex functions. The functions G_{ij} are assumed to be real-valued strictly increasing differentiable convex functions, defined on an open segment containing the range of f_{ij}. Note that in this case $G'_{ij}(t) > 0$ for all $t \in R$. As we have already pointed out, positive linear combinations of G-convex functions are not necessarily quasiconvex. For notational convenience, but without loss of generality, assume that each ψ_i is a sum of only two functions,

$$\psi_i = f_i + g_i,$$

such that the f_i are G_i-convex and the g_i are convex (i.e. x-convex).

Let us, therefore, restrict our attention to the following problem Q :

$$\min \phi(x) \tag{4.5}$$

subject to $x \in C$ and

(4.6) $$f_i(x)+g_i(x) \leq 0 \qquad i=1, \ldots, m.$$

Since our algorithm for solving Q is based on solving a sequence of convex programs, we shall first describe a technique to generate convex subsets of the nonconvex feasible set of Q. Since the G_i are assumed to be continuously differentiable convex functions we have for every t, $\hat{t}$

(4.7) $$G_i(t) \geq G_i(\hat{t})+G_i'(\hat{t})(t-\hat{t}).$$

Letting $t=f_i(x)$, $\hat{t}=f_i(\hat{x})$ we get

(4.8) $$G_i(f_i(x)) \geq G_i(f_i(\hat{x}))+G_i'(f_i(\hat{x}))(f_i(x)-f_i(\hat{x})).$$

Since $G'(t)>0$ we obtain after rearrangement :

(4.9) $$f_i(x) \leq \hat{f}_i(x, \hat{x}) = f_i(\hat{x})+[1/G_i'(f_i(\hat{x}))][G_i(f_i(x))-G_i(f_i(\hat{x}))].$$

As a special case of (4.9) let

(4.10) $$G(t) = \mathrm{e}^{rt}, \qquad r>0.$$

Then we get

(4.11) $$f_i(x) \leq f_i(\hat{x}) + \frac{1}{r}[\mathrm{e}^{rf_i(x)-rf_i(\hat{x})}-1].$$

If f_i is r-convex the right-hand side of (4.11) is the convex approximation of f_i. In [2] this inequality was proved by a special case of Young's inequality [5]. Letting $G(t)=t^a$, $a \geq 1$ we get

(4.12) $$f_i(x) \leq (1-1/a)\, f_i(\hat{x}) + \frac{f_i^a(x)}{a\, f_i^{a-1}(\hat{x})}.$$

Returning now to the general case let

(4.13) $$X(\hat{x}) = \{x : \hat{f}_i(x, \hat{x})+g_i(x) \leq 0, \qquad i=1, \ldots, m\}$$

and let $S(\hat{x}) = C \cap X(\hat{x})$. Then we have

THEOREM 4.1. For every $\hat{x} \in S$ the set $S(\hat{x})$ is a convex set and $\hat{x} \in S(\hat{x}) \subset S$.

Proof. By theorem 3.4 the functions $G_i f_i$ are convex, hence the $\hat{f}_i$ are convex functions. Then each of the $\hat{f}_i+g_i$ is convex and consequently $X(\hat{x})$ is a convex set. Since also C is convex we get that $S(\hat{x})$ is a convex set. If $\hat{x} \in S$ then $\hat{x} \in C$ and $\hat{x} \in X$. Since $\hat{f}_i(\hat{x}, \hat{x})=f_i(\hat{x})$ we have that $\hat{x} \in X(\hat{x})$ and

$\hat{x} \in S(\hat{x})$. Let $x \in S(\hat{x})$. Then, from (4.9) follows that $f_i(x) + g_i(x) \leq 0$, $i = 1, \ldots, m$, hence $x \in S$. QED.

As a consequence of Theorem 4.1 we see that the program $Q(x^k)$

$$\min \phi(x) \tag{4.14}$$

subject to $x \in C$ and

$$\hat{f}_i(x, x^k) + g_i(x) \leq 0 \qquad i = 1, \ldots, m \tag{4.15}$$

is a convex program provided $x^k \in S$. Note that $X(x^k)$ is the feasible set for program $Q(x^k)$.

The algorithm for solving Q will now be described. Starting from a feasible point $x^1 \in S$ we generate a sequence of feasible points $\{x^k\}$ as follows. For a given $x^k \in S$ solve the convex program $Q(x^k)$. Let x^{k+1} be any optimal solution of $Q(x^k)$. Under suitable regularity assumptions on Q every convergent subsequence of $\{x^k\}$ approaches a solution of Q. For proofs of convergence and further details the reader is referred to [12]. We conclude this study by relating the proposed algorithm to existing works. The idea of solving nonconvex programs having special structures by a sequence of convex programs is not new. A nonlinear program in which the constraint functions ψ_i are differentiable and concave is called a reverse-convex program, introduced by Rosen [11] and further extended by Meyer [10]. For a differentiable concave function the following relation holds for any two points x, $\hat{x}$ in its convex domain:

$$\psi(x) \leq \tilde{\psi}(x, \hat{x}) = \psi(\hat{x}) + (x - \hat{x})^T \nabla \psi(\hat{x}). \tag{4.16}$$

Replacing constraints $\psi_i(x) \leq 0$, $i = 1, \ldots m$ by

$$\tilde{\psi}_i(x, \hat{x}) \leq 0 \qquad i = 1, \ldots, m \tag{4.17}$$

one gets a sequence of convex programs whose feasible set is contained in the feasible set of the original problem. Note that the set of x satisfying (4.17) is a polyhedron since the $\tilde{\psi}$ are linear affine functions. If the set C, as defined in problem P above, is the whole space R^n, then a reverse-convex program can be solved by a sequence of linear programs. Consider now a nonconvex "generalized" geometric program, sometimes called "signomial" or "complementary" geometric program. Here the constraints are of the form

$$\frac{P(x)}{Q(x)} \leq 1 \tag{4.18}$$

where P and Q are posynomials. Let

(4.19) $$q_j(x) = c_j \prod_i x_i^{a_{ij}}$$

where $c_j > 0$ and $a_{ij} \in R$ are known constants.

Then Q can be written as

(4.20) $$Q(x) = \sum_j q_j(x).$$

It can be shown that

(4.21) $$Q(x) \geq \prod_j \left(\frac{Q(\hat{x})}{q_j(\hat{x})} q_j(x) \right)^{q_j(\hat{x})/Q(\hat{x})} = \tilde{Q}(x, \hat{x}),$$

where $\tilde{Q}$ is a one-term posynomial. Replacing constraints (4.18) by

(4.22) $$\frac{P(x)}{\tilde{Q}(x, \hat{x})} \leq 1,$$

where $\hat{x}$ satisfies (4.18), one gets an ordinary geometric program. It follows from (4.21) that any x satisfying (4.22) will also satisfy (4.18) and hence the feasible set of the ordinary geometric program is contained in the feasible set of the generalized geometric program. Constructing a sequence of points $\{x^k\}$ in the same manner as for problem P above it can be seen that a generalized geometric program can be solved by a sequence of ordinary geometric programs [3, 4].

The feasible sets of ordinary geometric programs can be easily transformed into convex sets thus we again have the case of solving a nonconvex program by a sequence of convex programs. The algorithm proposed in this work extends this idea by using the notion of G-convexity.

References

1. Avriel, M. r-Convex functions. *Mathematical Programming* (forth-coming).
2. Avriel, M. Solutions of certain nonlinear programs involving r-convex functions. *Journal of Optimization Theory and Applications* (forthcoming).
3. Avriel, M. and Williams, A. C. Complementary geometric programming. *S.I.A.M. Journal of Applied Mathematics*, **19**: 125–141, 1970.
4. Avriel, M. and Williams, A. C. An extension of geometric programming with applications in engineering optimization. *Journal of Engineering Mathematics*, **5**: 187–194, 1971.
5. Hardy, G. H., Littlewood, J. E. and Polya, G. *Inequalities*. Cambridge University Press, Cambridge, 1967.

6. KLINGER, A. and MANGASARIAN, O. L. Logarithmic convexity and geometric programming. *Journal of Mathematical Analysis and Applications*, **24**: 388–408, 1968.
7. MACLANE, S. and BIRKHOFF, G. *Algebra*. Macmillan, London, 1967.
8. MANGASARIAN, O. L. Pseudo-convex functions. *Journal of S.I.A.M. Control*, Ser. A. **3**: 281–290, 1965.
9. MARTOS, B. Nem-Lineáris Programmozási Módszerek Hatôköre (The power of nonlinear programming methods). *MTA Közgazdaságtudományi Intézetének*. Közleményei No. 20, Budapest, 1966 (in Hungarian).
10. MEYER, R. R. The Solution of Non-Convex Optimization Problems by Iterative Convex Programming. *Ph. D. Thesis*, University of Wisconsin, 1968.
11. ROSEN, J. B. Iterative solution of nonlinear optimal control problems. *Journal of S.I.A.M. Control*, **4**: 223–244, 1966.
12. ZANG, I. Generalized Convex Functions and Programming. *D. Sc. Thesis*, Technion, Israel Institute of Technology.

CHAPTER 3

A GENERAL CONVERGENCE THEOREM AND ITS APPLICATION TO A PENALTY ALGORITHM

FRANS VAN WINCKEL

Summary

The classical convergence theorems do not provide a straightforward approach to proving algorithmic convergence in a number of cases.

In this paper we try to derive a more general convergence theorem characterized by weaker conditions. This theorem is then applied to a penalty algorithm transforming, under adequate concavity conditions, a constrained Mathematical-Programming problem into an unconstrained one.

1. Algorithm

Consider the general programming problem defined as:

$$\max_{X \in E^n} f(x)$$

subject to

(1.1) $$g_i(x) \geq 0 \qquad i = 1, 2, \ldots, m.$$

For all $k \geq 1$ define a set $V_k \subset V$ and the algorithmic point-to-set map

$$A_k : V_k \to V_{k+1}.$$

We call an algorithm the sequence of points $\{z^k\}_{\mathscr{K}}$ generated, where $\mathscr{K}$ is an infinite set of positive integers. Observe that we use the variable z instead of the variable x.

Frequently z is actually the variable of the programming problem and in this case $z = x$. More generally however both variables are different. If

$$A_k(z^k) = \emptyset$$

the procedure stops. Otherwise, select the successor point by

$$z^{k+1} \in A_k(z^k).$$

It follows from this definition that any sequence $\{y^k\}_1^\infty$ in V can be interpreted as an algorithm. Define the sets

$$V_k = [y^k] \text{ for all } k,$$

where $[y^k]$ indicates the set consisting of the point y^k. Let $z^1 = y^1$ and

$$A_k(z^k) = [y^{k+1}] \text{ for all } k.$$

2. Convergent Algorithm

Given a problem (1.1) and a solution set $S \subset V$, a convergent algorithm is an algorithm with the following properties:

(2.1) $$A_k(z^k) = \emptyset \rightarrow z^k \in S \vee S = \emptyset$$

(2.2) $$z^k \in S \rightarrow A_k(z^k) = \emptyset \vee A_k(z^k) \subset S$$

(2.3) $$\{z^k\}_{\mathscr{K}} : \forall k : z^k \notin S \rightarrow (\forall k : z^k \in X \subset V \text{ and } X \text{ not compact} \rightarrow S = \emptyset) \vee (\forall k : z^k \in X \subset V \text{ and } X \text{ compact} \rightarrow \{z^k\}_{\mathscr{K}^1}, z^k \rightarrow z^\infty : z^\infty \in S)$$

where $\mathscr{K}^1 \subset \mathscr{K}$ is an infinite set of positive integers, and z^∞ the limit value of a sequence.

3. Classic Convergence Theorem

Let $A : V \rightarrow V$ determine the algorithm $\{z_k\}_{\mathscr{K}}$, and $S \subset V$ be given. The algorithm converges if:

1. $\forall k : z^k \in X \subset V$ and X is compact,
2. $Z : V \rightarrow E^1$ and Z is continuous
   a) $z^k \notin S \rightarrow \forall y \in A(z) : Z(y) > Z(z)$,
   b) $z^k \in S \rightarrow (A(z^k) = \emptyset) \vee y \in A(z^k) : Z(y) \geq Z(z)$,
3. $z^k \notin S \rightarrow A$ is closed at z.

The classic convergence theorem can be used to prove the convergence of numerous algorithms.

Condition 1 guarantees that all sequences are well-behaved. Without this condition a subsequence might have a limit outside the set V or might diverge. Note however that the feasible set F need not be compact.

Condition 2a requires improvement at each iteration of the algorithm until a solution point is reached. We may interpret the Z function as an adaptation function that indicates the algorithm's progress. Many algorithms use the objective function f to measure their progress, and in that case $Z = f$. Some algorithms, however, solve the M.P. problem by transforming it into another problem, and the function Z can then be considered as the objective function for the transformed problem. Condition 2b is similar to condition 2 of a convergent algorithm.

Condition 3 is required to prohibit the discontinuities that may cause nonconvergence. Observe however that closedness of the map is not required at a solution point.

In establishing convergence, the proof of condition 3 is usually the most challenging.

4. Lemma 1

$$\text{Let } h(z) = \inf\{|z-y|;\ y \in S\}.$$

Then

1. $h(z)$ is continuous,
2. S is closed $\rightarrow (h(z) = 0 \leftrightarrow z \in S)$.

Proof. Consider

$$\{z^k\}_1^\infty. \tag{4.1}$$

$$y^1 \in S : |z^k - y^1| \leq |z^\infty - y^1| + |z^\infty - z^k|$$

$$\rightarrow h(z^k) = \inf\{|z^k - y|;\ y \in S\} \leq |z^\infty - y^1| + |z^\infty - z^k|$$

$$\rightarrow h(z^k) \leq \inf\{|z^\infty - y|;\ y \in S\} + |z^\infty - z^k|$$

$$\rightarrow h(z^k) \leq h(z^\infty) + |z^\infty - z^k|. \tag{4.2}$$

Similarly

$$h(\infty) \leq h(z^k) + |z^\infty - z^k| \tag{4.3}$$

and from (4.2) and (4.3)

$$h(z^k) - h(z^\infty) \leq |z^\infty - z^k|. \tag{4.4}$$

Exploiting (4.1)

$$z^k \to z^\infty \to k \geq P : |z^\infty - z^k| < \varepsilon$$
$$\to k \geq P : |h(z^k) - h(z^\infty)| < \varepsilon$$
$$\to \lim_{k\to\infty} h(z^k) = h(z^\infty),$$

which proves continuity.

By the definition of $h(z)$

$$z \in S \to h(z) = 0$$

and by the definition of an infimum and the fact that S is a closed set

$$h(z) = 0 \to z \in S.$$

5. Lemma 2

Let $\{z^k\}_1^\infty \in X \subset V$ and X is compact.

$$\{z^k\}_{\mathscr{K}} \quad \text{and} \quad z^\infty \in S \to \lim_{k\to\infty} h(z^k) = 0.$$

Proof. X is compact $\to \{z^k\}_{\mathscr{K}} : z^k \to z^\infty,\ k \in \mathscr{K}$ and

$z^\infty \in S \to h(z^\infty) = 0$ and

h is continuous $\to \lim_{k\in\mathscr{K}} h(z^k) = h(z^\infty) = 0$, and

$\mathscr{K}$ is an arbitrary subsequence $\to \lim_{k\to\infty} h(z^k) = h(z^\infty) = 0.$

6. Lemma 3

Suppose $Z : V \to E^1$ is continuous.

$$\forall z^k : L_k \in I^+ \wedge l \geq L_k + k \wedge Z(z^l) \geq Z(z^k)$$
$$\wedge \{z^k\}_{\mathscr{K}} \quad \text{and} \quad z^\infty \in S \to \lim_{k\to\infty} Z(z^k) = Z(z^\infty)$$

where I^+ is the set of positive integers.

Proof. For simplicity we define $Z^k = Z(z^k)$, $\forall k$. By continuity of Z

(6.1) $$\lim_{k\in\mathscr{K}} Z^k = Z^\infty,$$

(6.2) $$\varepsilon < 0 \wedge k', k'' > P \wedge k'k'' \in \mathscr{K} \to |z^{k'} - z^{k''}| < \varepsilon.$$

$P^1 = L_{k'} + k' \wedge l > P^1 \rightarrow Z^{k'} \leq Z^l$

$P^2 = L_l + l \wedge k'' > P^2 \wedge k'' \in \mathscr{K} \rightarrow Z^l \leq Z^{k''}$.

From (6.2) $\rightarrow |Z^l - Z^{k'}| \leq |Z^{k'} - Z^{k''}| < \varepsilon \xrightarrow{\text{limit}} |Z^l - Z^{\infty}| < \varepsilon, \varepsilon > 0, l > P_1$.

7. General Convergence Theorem

An algorithm converges if:

1. $A_k(z^k) = \emptyset \rightarrow z^k \in S \vee S = \emptyset$
2. $z^k \in S \rightarrow A_k(z^k) = \emptyset \vee A_k(z^k) \subset S$
3. $\{z^k\}_{\mathscr{K}}, \forall k : z^k \notin S \rightarrow (S \neq \emptyset \rightarrow z^k \in X \subset V$ and X is compact)
4. $\{z^k\}_{\mathscr{K}}, \forall k : z^k \in X \subset V$ and X compact $\rightarrow Z : X \rightarrow E^1$ and Z continuous:
   a) $(z^k, L_k \in N_0) : (l \geqq k + L_k \rightarrow Z(z^l) \geqq Z(z^k))$
   b) $(\{z^k\}_{\mathscr{K}^1}, z^k \rightarrow z', \mathscr{K}^1 \subset \mathscr{K}, k \in \mathscr{K}^1, z' \notin S) \rightarrow$
   $(\exists k' : Z(z^{k'}) > Z(z')) \vee (\{(z^k)\}_{\mathscr{K}^2}, z^k \rightarrow z'', \mathscr{K}^2 \subset \mathscr{K}^1,$
   $k \in \mathscr{K}^2 : Z(z'') > Z(z'))$.

These four conditions are sufficient for convergence; they are necessary if S is closed.

Proof.

Sufficiency. Conditions 1 and 2 are similar to conditions 1 and 2 of a convergent algorithm. Conditions 3 and 4 consider the case of an infinite sequence that does not own a solution point. Condition 3 requires the existence of a compact set X. If all points are in a compact set, it only remains to prove that the limit of any convergent subsequence is a solution.

Suppose

$$\{z^k\}_{\mathscr{K}^1}, z^k \rightarrow z', \mathscr{K}^1 \subset \mathscr{K}, k \in \mathscr{K}^1 \text{ and } z' \notin S.$$

From 4 and lemma 3 one gets

$$\lim_{k \rightarrow \infty} Z(z^k) = Z(z'). \tag{7.1}$$

Using compactness of X we may find

$\mathscr{K}^2 \subset \mathscr{K}^1, z^k \rightarrow z'', \lim_{k \in \mathscr{K}^2} Z(z^k) = Z(z'')$, and by lemma 3, $\lim_{k \rightarrow \infty} Z(z^k) = Z(z'')$.

Consequently $Z(z') = Z(z'')$, which is in contradiction with condition 4.

Necessity. Again conditions 1 and 2 of the algorithm and of the definition are similar.

By the convergence definition, if an infinite sequence of points none of which is a solution is generated and if a solution exists, all points are in a compact set X. Hence condition 3 holds.

It remains to prove condition 4 of the convergence theorem

$$Z(z) = -\inf\{|z-y|;\ y \in S\} = -h(z).$$

From lemma 1 it follows that $Z: V \to E^1$ is continuous and $Z(z)=0$ if $z \in S$ and S closed.

Consider $\{Z(z^k)\}_1^\infty$. Observe

a) $Z = -h$

b) $\forall z_k \in X$

c) $\{z^k\}_{\mathscr{K}},\ z^k \to z^\infty \to z^\infty \in S \to Z(z^\infty) = 0$

(7.2) d) $\lim\limits_{k\to\infty} Z(z^k) = 0$

Since by assumption $z^k \notin S \to Z(z^k) < 0$ and from (7.2)

$$Z(z^k) \le Z(z^1) < 0$$

which proves 4. The other part of 4 holds as the limit of any convergent subsequence must be a solution.

8. Comparison between the Classical and the General Convergence Theorem

Four difficulties of the classic convergence theorem are immediately evident:

1. requirement that the algorithmic map be a closed map;
2. requirement of strict improvement of the Z function if z^k is not a solution;
3. the compactness assumption;
4. the requirement that the map A is defined on the whole space V.

The closed-map condition in the classic convergence theorem was required to guarantee convergence at a point which is not a solution.

Suppose

$$\{z^k\}_{\mathscr{K}^1},\ z^k \to z',\ z' \notin S.$$

By the classic convergence theorem, there must exist another

$$\{z^{k+1}\}_{\mathscr{K}^1},\ \mathscr{K}^1 \subset \mathscr{K},\ z^{k+1} \to z''$$

and closedness ensures that

(8.1) $$z'' \in A(z') \to Z(z'') > Z(z').$$

A similar convergence condition is found back in 4 of the general convergence theorem. The basic assumption however is weaker.

The strict adaptation requirement of the Z function is weakened by requiring improvement only after a finite number of iterations:

$$Z(z^l) \geq Z(z^k)$$

for all $l \geq L_k + k$ where L_k may depend upon k and Z^k.

Some weaker compactness assumption remains. The algorithmic map previously specified $A: V \to V$, was defined on all of the space V, and was independent of z^k. Therefore A_k, defined only on a portion of V, say $V_k \subset V$, and dependent on z^k, is introduced.

9. Penalty Methods

Penalty methods seek the solution of the M.P. problem by transforming it into a sequence of problems without constraints.

Consider the function

$$P^1(x) = \begin{cases} 0 & x \in F \\ -L & x \notin F \end{cases}$$

where F is the feasible set.

We form

(9.1) $$C(x) = f(x) + P^1(x).$$

Clearly $P^1(x)$ produces a sufficiently high penalty, L, for leaving the feasible region, and x^* is optimal for (9.1), if and only if, it is optimal for (1.1).

Hence we solve

(9.2) $$\max_{x \in E^n} C(x).$$

Regrettably (9.1) is unsolvable in a practical sense, due to its severe discontinuities, very large L and resultant rounding error in the computer.

Therefore consider the penalty function

(9.3) $$P(x) = \frac{1}{r^k} P^2(x)$$

with:

(9.4)
$$\begin{cases} \{r^k\}_1^\infty, \ r^k \to 0, \ r^k > r^{k+1} \\ P^2(x) = -\sum_i \min\,[g_i(x), 0]^2 & \begin{cases} = 0 & \text{if} \quad x \in F \\ < 0 & \text{if} \quad x \notin F \end{cases} \end{cases}$$

such that

$$C(X, r^k) = f(x) + \frac{1}{r^k} P^2(x).$$

We generate the sequence $\{x^k\}_1^\infty$ and, as will be proved, a limit of any convergent subsequence is an optimal point for the problem.

10. Lemma 4

We assume that f and g_i are continuous and that an optimal point exists. Suppose also that $x^k \in X$ and X is compact.

1. $C(x^k, r^k) \geq C(x^{k+1}, r^{k+1})$.
2. $P(x^k) \leq P(x^{k+1})$.
3. $f(x^k) \leq f(x^{k+1})$.

Proof.

1. From (9.3) $\to P(x) \leq 0, \forall x$.

Because $\dfrac{1}{r^k} < \dfrac{1}{r^{k+1}}$

$$f(x^{k+1}) + \frac{1}{r^k} P(x^{k+1}) \geq f(x^{k+1}) + \frac{1}{r^{k+1}} P(x^{k+1}).$$

Furthermore

$$C(x^k, r^k) = f(x^k) + \frac{1}{r^k} P(x^k) \geq f(x^{k+1}) + \frac{1}{r^k} P(x^{k+1})$$

$$\downarrow \qquad \geq f(x^{k+1}) + \frac{1}{r^{k+1}} P(x^{k+1})$$

$$C(x^k, r^k) \geq C(x^{k+1}, r^{k+1}).$$

2. (10.1)
$$\begin{cases} f(x^k)+\frac{1}{r^{k+1}}P(x^k)\le f(x^{k+1})+\frac{1}{r^{k+1}}P(x^{k+1}) \\ f(x^{k+1})+\frac{1}{r^k}P(x^{k+1})\le f(x^k)+\frac{1}{r^k}P(x^k) \end{cases}$$

$$\to \frac{1}{r^{k+1}}[P(x^k)-P(x^{k+1})]\le \frac{1}{r^k}[P(x^k)-P(x^{k+1})]$$

$$\to P(x^k)\le P(x^{k+1}).$$

3. By (10.1)

$$f(x^{k+1})-f(x^k)\le \frac{1}{r^k}[P(x^k)-P(x^{k+1})]\le 0.$$

11. Convergence Proof of the Penalty Algorithm

Consider the general programming problem (1.1) and suppose $z=x$; f and g_i are continuous, $S\neq\emptyset$; $\{x^k\}_1^\infty\in X$ and X is compact. Then $\{x^k\}_1^\infty$ converges to a solution point.

Proof. If x^* is termed a solution, then clearly conditions 1 and 2 of the general convergence algorithm hold. Condition 3 holds as X is compact. To prove condition 4 we define $Z(x)=-f(x)$. From lemma 4: $Z(x^{k+1})\geqq Z(x^k)$. Put $L_k=1$ and 4a) holds in section 7.

Let $\{x^k\}_{\mathscr{K}}$, $x^k\to x'$, $k\in\mathscr{K}$. By (9.4) one gets:

(11.1) $$x^*\in F \text{ implies } f(x^*)=C(x^*,r^k)\leqq C(x^k,r^k)=f(x^k)+\frac{1}{r^k}P(x^k).$$

From lemma 4 and (11.1) the sequence $\{C(x^k,r^k)\}_1^\infty$ is nonincreasing and bounded below by $f(x^*)$. Hence it has a limit:

$$C^\infty=\lim_{k\to\infty}C(x^k,r^k)=\lim_{k\in\mathscr{K}}C(x^k,r^k).$$

From the continuity of f we find that $\lim_{k\in\mathscr{K}}f(x^k)=f(x^1)$, and using (11.1),

$$\lim_{k\in\mathscr{K}}\frac{1}{r^k}P(x^k)=\lim_{k\in\mathscr{K}}[C(x^k,r^k)-f(x^k)]=C^\infty-f(x^1)$$

so that

(11.2) $$\lim_{k \in \mathscr{K}} P(x^k) = 0 \rightarrow P(x^1) = 0 \rightarrow x^1 \in F\,.$$

By lemma 4, $f(x^k) \geqq f(x^*)$ so that

(11.3) $$\lim_{k \in \mathscr{K}} f(x^k) = f(x^1) \geqq f(x^*)\,.$$

Combining (11.2) and (11.3) yields $(x^1 \in S \wedge f(x^1) \geqq f(x^*))$ so that x^1 must be a solution and condition 4*b*) holds. QED.

References

1. Fiacco, A. V. and McCormick, G. P. Computational algorithm for the sequential unconstrained minimization technique for nonlinear programming. *Management Science*, **10**: 601–617, 1964.
2. Mangasarian, O. L. Necessary optimality conditions in the presence of equality and inequality constraints. *Journal of Mathematical Analysis and Applications*, **17**: 37–47, 1965.
3. Stong, R. E. A note on the sequential unconstrained minimization technique for nonlinear programming. *Management Science*, **12**: 142–144, 1966.
4. Zangwill, W. I. Nonlinear programming via penalty functions. *Management Science*, **13**: 344–357, 1967.

CHAPTER 4

SIGNOMIAL GEOMETRIC PROGRAMMING: DETERMINING THE GLOBAL MINIMUM

URY PASSY

Summary

We present a method for determining the global minimum of signomial programs, together with a numerical example. It is shown that a local minimum is in many cases not sufficient and the extra computational effort, required for locating the global minimum, is worthwhile.

1. Introduction

Many efficient algorithms for minimizing convex programs were developed in the last decade. However, many problems, especially in engineering design, are not convex. Not much has been done for solving such programs—except, of course, in the case of integer programming. An algorithm for nonconvex quadratic programs is described by Cottle and Mylander [4], a method for solving a certain class of nonconvex separable functions is described by Falk and Soland [7].

Signomial programs [6] are not convex. Methods for transforming such programs into a sequence of geometric programs were described in [1], [12]. But, the "solution" is only a pseudominimum [10], i.e. a point satisfying the necessary conditions for a minimum. In most practical cases a pseudominimum will be a local minimum [1].

Finding the global minimum in nonconvex programs will usually involve massive computations. It is uncertain whether the extra effort required to locate a global minimum, as compared with a "good solution", actually justifies itself. For example, Cooper [3] has claimed that the minimum of the nonconvex location allocation problem is not significantly better than his "solution". The same was claimed by Nemhauser [2] for the scheduling problem.

The nonconvex constraint set for signomial programs can be described by a union of finite convex subsets. The minimum of the objective function over each convex subset may be a local minimum for the original problem (see numerical example). Therefore, such programs have in general many local minima. Moreover, the feasible region is not necessarily connected. The constraint set described by the following three inequalities :

$$1 \geq e^{-1}x \quad 1 \geq x^{-1} \quad 1 \leq e(e+1)^{-1}x^{-1}+(e+1)^{-1}x$$

consists of two points $\{1, e\}$. It is therefore felt that for such programs a local minimum is not a good "solution", and the extra effort is justified.

In the present proposed method the original feasible region is replaced by a union of convex sets, and the original program is transformed into a geometric program with conditional constraints [8]. Such programs are equivalent to 0–1 mixed geometric programs and in some cases to 0–1 mixed linear programs. (See Appendix A2.) The method suggested can be practically used, therefore, only on small problems.

Following Zoutendijk there are two types of nonlinear problems : the first one is the economic-type problem which is large and almost linear; the second is the engineering-design type which is small and highly nonlinear. Signomial programs appear in engineering-design problems and the new algorithm can successfully be applied to these problems.

It is felt that the method suggested, implicit enumeration, becomes more efficient as the difference between the global minimum and the local one increases. (See numerical example.)

2. Signomial Optimization

There are two equivalent formulations for signomial optimization.

A. First formulation

(2.1) $$\operatorname{Min} p_0(\mathbf{x})$$

over the domain defined by

(2.2) $$p_k(\mathbf{x})-q_k(\mathbf{x}) \leq 1 \qquad i=1, \ldots, P$$

(2.3) $$\mathbf{x} \in (R^m)^+ .$$

The functions p_k and q_k are posynomials [6] defined by

(2.4a) $$p_k(\mathbf{x}) = \sum_{i=1}^{I_1(k)} C_{ik} \prod_{j=1}^{m} x_j^{\alpha_{ijk}}$$

(2.4b) $$q_k(\mathbf{x}) = \sum_{i=1}^{I_2(k)} D_{ik} \prod_{j=1}^{m} x_j^{\beta_{ijk}} .$$

The coefficients $\underline{C}_{ik}$ and $\underline{D}_{ik}$ are nonnegative, $\underline{\alpha}_{ijk}$ and $\underline{\beta}_{ijk}$ are given real numbers and $I_i(0), \ldots, I_i(P)$ $(i=1, 2)$ are given integers $(I_2(0)=0)$.

B. Second formulation

$$\text{Min } g_0(\mathbf{x}) \tag{2.5}$$

over the domain defined by

$$g_k(\mathbf{x}) \leq 1 \qquad k=1, \ldots, P_1 \tag{2.6a}$$

$$h_k(\mathbf{x}) \geq 1 \qquad k=1, \ldots, P_2 \tag{2.6b}$$

$$\mathbf{x} \in (R^l)^+ . \tag{2.7}$$

The functions $\underline{h}_k$ and $\underline{g}_k$ are posynomials.

Representation A can be transformed into B using the following relations:

$$g_k(x_1, \ldots, x_m, x_{m+1}, \ldots, x_l) = \begin{cases} p_k(x_1, \ldots, x_m) & \text{if } p_k \text{ or } q_k \\ \text{are the zero function} & \\ x_{m+k}^{-1}\, p_k(x_1, \ldots, x_m) & \end{cases} \tag{2.8a}$$

$$h_k(x_1, \ldots, x_m, x_{m+1}, \ldots, x_l) = \begin{cases} 2+q_k(x_1, \ldots, x_m) & \text{if } p_k=0 \\ 0 \quad \text{if } q_k=0 & \\ x_{m+k}^{-1}\, q_k(x_1, \ldots, x_m)+x_{m+k}^{-1} . & \end{cases} \tag{2.8b}$$

The reverse transformation is given by

$$\begin{aligned} p_k(\mathbf{x}) &= \begin{cases} g_k(\mathbf{x}) & k=1, \ldots, P_1 \\ 2 & k=P_1+1, \ldots, P_1+P_2 \end{cases} \\ q_k(\mathbf{x}) &= \begin{cases} 0 & k=1, \ldots, P_1 \\ h_{k-P_1}(\mathbf{x}) & k=P_1+1, \ldots, P_1+P_2 . \end{cases} \end{aligned} \tag{2.9}$$

If $P_2=0$ the problem is a regular geometric program and any local minimum is also a global minimum. If $P_2>0$ the program is usually nonconvex and is called signomial program.

Presently the second formulation will be used. The point set defined by

(2.6a), (2.7) will be denoted by $\underset{\sim}{S}$ and the problem is therefore

$$\min g_0(\mathbf{x}) \tag{2.10}$$

over the domain $D \cap S$ where $\underset{\sim}{D}$ is defined by (2.6b).

3. Expansion of Signomial Constraints

Let $\underset{\sim}{D}$ be the point set defined by

$$D = \{\mathbf{x} / \sum_{i=1}^{I} C_i \prod_{j=1}^{m} x_j^{\alpha_{ij}} \geq 1;\ \mathbf{x} \in (R^m)^+\} \tag{3.1}$$

and let $D(\mathbf{U})$ be the point set defined by

$$D(\mathbf{U}) = \{\mathbf{x} / \operatorname*{Max}_{1 \leq i \leq I} \{C_i \prod_{j=1}^{m} x_j^{\alpha_{ij}}/u_i\} \geq 1;\ \mathbf{x} \in (R^m)^+\}. \tag{3.2}$$

The vector $\mathbf{U}$ is any positive vector. The set of all normal positive vectors will be denoted by

$$\mathscr{U} = \{\mathbf{U} / \sum_{i=1}^{I} u_i = 1;\ u_i > 0\}.$$

LEMMA 3.1. The following two sets are identical:

$$D = \bigcap_{\mathbf{U} \in \mathscr{U}} D(\mathbf{U}).$$

Proof. (a) Let $\mathbf{x} \in D$ then

$$\sum_{i=1}^{I} C_i \prod_{j=1}^{m} x_j^{\alpha_{ij}} \geq 1.$$

From the generalized mean inequality,

$$1 \leq \sum_{i=1}^{I} u_i (C_i \prod_{j=1}^{m} x_j^{\alpha_{ij}}/u_i) \leq \max_{1 \leq i \leq I} (C_i \prod_{j=1}^{m} x_j^{\alpha_{ij}}/u_i) \qquad \forall \mathbf{U} \in \mathscr{U}.$$

Therefore, $\mathbf{x} \in D(\mathbf{U}) \quad \forall \mathbf{U} \in \mathscr{U}$.

(b) Let $\mathbf{x} \in \bigcap_{\mathbf{U} \in \mathscr{U}} D(\mathbf{U})$ and $\mathbf{x} \notin D$, i.e.

$$\sum_{i=1}^{I} C_i \prod_{j=1}^{m} x_j^{\alpha_{ij}} < 1.$$

Choose

$$u_i^* = \frac{C_i \prod_{j=1}^{m} x_j^{\alpha_{ij}}}{\sum_{k=1}^{I} C_k \prod_{j=1}^{m} x_j^{\alpha_{kj}}} \qquad i = 1, \ldots, I.$$

Clearly $\mathbf{U}^* \in \mathscr{U}$ and

$$\max_{1 \le i \le I} \{C_i \prod_{j=1}^{m} x_j^{\alpha_{ij}}/u_i^*\} = \sum_{k=1}^{I} C_k \prod_{j=1}^{m} x_j^{\alpha_{kj}} < 1.$$

Hence $\mathbf{x} \notin \bigcap_{\mathbf{U} \in \mathscr{U}} D(\mathbf{U})$, which contradicts the initial assumption. QED.

Given a signomial program[1] (called S.P.)

(3.3) $$\min x_1$$

subject to

(3.4a) $$g_k(\mathbf{x}) \le 1 \qquad k = 1, \ldots, P_1$$

(3.4b) $$\min_{1 \le k \le P_2} \{h_k(\mathbf{x})\} \ge 1$$

$$\mathbf{x} \in \{R^m\}^+.$$

Let E.S.P.(**U**) be the following program:

(3.5) $$\min x_1$$

subject to

(3.6a) $$g_k(\mathbf{x}) \le 1 \qquad k = 1, \ldots, P_1$$

(3.6b) $$\max \{D_{ik} \prod_{j=1}^{m} x_j^{\beta_{ijk}}/u_{ik}\} \ge 1 \quad \text{if} \quad h_k(\mathbf{x}) = \min_{1 \le j \le P_2} h_j(\mathbf{x}).$$

The vector $\mathbf{U} = [\mathbf{U}_1, \mathbf{U}_2, \ldots, \mathbf{U}_{p_2}]$ is an $\sum_{i=1}^{P_2} I(i)$ vector, and $\mathbf{U}_k$ is a given positive vector. If $\mathbf{U}_k$ satisfies the normality condition,

(3.7) $$\sum_{i=1}^{I(k)} u_{ik} = 1 \qquad u_{ik} > 0,$$

the E.S.P.(**U**) is an extension of S.P.

[1] There is no loss of generality by assuming that $g_0(\mathbf{x}) = x_1$ [9].

4. The Method of Solution

The method belongs to the "Cutting Plane" family. Since the problem is not convex a surface replaces the hyperplane cut. A description of the method together with the convergence proof is given below. The solution scheme is described in Sec. 5.

It is assumed that the constraint set (3.7) can be imbedded in a compact convex set $D \in (R^m)^+$. Let $\underline{D}_0 = D \cap S$.

The Method

Step 0: Minimize (3.5) over D_0.

Step k: (i) Let $h_j(\mathbf{x}^{(k-1)}) = \min_{1 \le j \le P_2} h_j(\mathbf{x}^{(k-1)})$.

Evaluate $\mathbf{U}_J^{(k)}$ by the formula

$$u_{iJ}^{(k)} = D_{iJ} \prod_{l=1}^{m} (x_l^{(k-1)})^{\beta_{ilJ}} / h_J(\mathbf{x}^{(k-1)}). \tag{4.1}$$

(ii)Minimize (3.5) over

$$D^{(k)} = D^{(k-1)} \cap D(\mathbf{U}_J^{(k)}). \tag{4.2}$$

The algorithm terminates when

$$\min_{1 \le j \le P_2} h_j(\mathbf{x}^{(k)}) \ge 1 - \varepsilon .$$

The cut after each iteration excludes $\mathbf{x}^{(k-1)}$ and its neighborhood from the feasible set. This follows from the following arguments [11]:

$$\max_{1 \le i \le I(J)} \{D_{iJ} \prod_{l=1}^{m} (x_l^{(k-1)})^{\beta_{ilJ}} / u_{iJ}^{(k)}\} = h_J(\mathbf{x}^{(k-1)}). \tag{4.3}$$

If $\mathbf{x}^{(k-1)}$ is not feasible, i.e. $h_J(\mathbf{x}^{(k-1)}) < 1$ then

$$\max_{1 \le i \le I(J)} \{D_{iJ} \prod_{l=1}^{m} (x_l^{(k-1)})^{\beta_{ilJ}} / u_{iJ}^{(k)}\} < 1 \tag{4.4}$$

since

$$h_J(x) = \min h_j(x) .$$

Therefore, $\mathbf{x}^{(k-1)} \notin D^{(k)}$.

The sets $\underline{D}^{(k)}$ generate a sequence $\{D^{(k)}\}$ with $D^{(k)} \subset D^{(k-1)}$. Thus, $\mathbf{x}_1^{(k)} \ge \mathbf{x}_1^{(k-1)}$ so $\{\mathbf{x}_1^{(k)}\}$ is a bounded monotonic sequence. Since $\underline{D}$ is a compact subset of $(R^m)^+$ there is a convergent subsequence $\{\mathbf{x}^{(k)}\}_{k \in K} \to \mathbf{x}^*$. It remains to show that each convergent subsequence will satisfy (3.4b) in the limit.

LEMMA 4.1. Let $\{\mathbf{x}^{(k)}\}_{k\in K}\to\mathbf{x}^*$ be a convergent subsequence, then

$$h_j(\mathbf{x}^*)\geq 1 \qquad \forall 1\leq j\leq P_2.$$

Proof. Define the following P_2 sequences:

$$K(j)=\{k/h_j(\mathbf{x}^{(k)})=\min_{1\leq l\leq P_2} h_l(\mathbf{x}^{(k)})\}.$$

(i) Since (4.1) is continuous

$$\{\mathbf{U}_j^{(k)}\}_{k\in K(j)}\to\mathbf{U}_j^* \qquad j=1,\ldots,P_2.$$

(ii) Let

$$L1=\{j/K(j)\text{ is finite}\}$$
$$L2=\{j/K(j)\text{ is infinite}\}.$$

(a) If $j\in L1$ then $\exists N$ such that

$$h_j(\mathbf{x}^{(k)})\geq\min_{i\in L2} h_i(\mathbf{x}^{(k)}) \qquad \forall k>N. \tag{4.5}$$

This follows from step (i) of the method described.
(b) If $j\in L2$ then

$$\lim_{\substack{k\in K(j)\\ k\to\infty}}\ \max_{1\leq i\leq I(j)}\{D_{ij}\prod_{l=1}^{m}(x_l^{(k)})^{\beta_{ilj}}/u_{ij}^{(k)}\}=D_{ij}\prod_{l=1}^{m}(x_l^*)^{\beta_{ilj}}/u_{ij}^*\geq 1 \qquad \forall 1\leq i\leq I(j). \tag{4.6}$$

(iii) From the generalized mean inequality it follows that [11], if

$$D_{ij}\prod_{l=1}^{m}(x_l^*)^{\beta_{ilj}}/u_{ij}=D_{sj}\prod_{l=1}^{m}(x_l^*)^{\beta_{slj}}/u_{sj}^* \qquad \forall s,j \tag{4.7}$$

then

$$h_j(\mathbf{x}^*)=D_{ij}\prod_{l=1}^{m}(x_l^*)^{\beta_{ilj}}/u_{ij}^*. \tag{4.8}$$

Therefore, $h_j(\mathbf{x}^*)\geq 1 \quad \forall j\in L2$ (from (4.6), (4.8)); also

$$h_j(\mathbf{x}^*)\geq\min_{i\in L2} h_i(\mathbf{x}^*)\geq 1.\ \text{QED.}$$

5. The Algorithm

Consider the following point set

$$X=\{\mathbf{x}/\mathbf{x}\in S;\quad \max_{1\leq i\leq n}\{D_i\prod_{j=1}^{m}x_j^{\beta_{ij}}/u_i\}\geq 1\}.$$

The above set is equivalent to X_1 which is defined by the following condition:

$$X_1 = \{\mathbf{x}/\mathbf{x} \in S; \quad x_j = e^{Z_j}; \quad \sum_{j=1}^{m} \beta_{ij} Z_j \geq \ln u_i/D_i - M_i y_i; \quad i = 1, \ldots, n\},$$

where M_i satisfies

$$M_i + \min_{\mathbf{x} \in S} \sum_{j=1}^{m} \beta_{ij} \ln x_j \geq \ln u_i/D_i .$$

The variables y_i are defined by

$$\sum_{i=1}^{n} y_i \leq 1 \qquad y_i \in \{0\ 1\} .$$

Let $Y_i = e^{y_i}$; in this case, X is equivalent to X_2 which is defined by the following conditions:

$$X_2 = \{\mathbf{x}/\mathbf{x} \in S; \ U_i/D_i \prod_{j=1}^{m} x_j^{-\beta_{ij}} Y_i^{-M_1} \leq 1, \ i = 1, \ldots, n; \ \prod_{i=1}^{n} Y_i \leq e, \ Y_i \in \{1, e\}\}.$$

The constraints defining X_2 are posynomial functions. Thus, an E.S.P. (**U**) ((3.5), (3.6)) is equivalent to the following geometric program:

(5.1) $$\min x_1$$

over the constraint set

(5.2) $$g_k(\mathbf{x}) \leq 1 \qquad k = 1, \ldots, P_1$$

(5.3) $$u_{ik}/D_{ik} \prod_{j=1}^{m} x_j^{-\beta_{ijk}} Y_{ik}^{-M_{ik}} \leq 1 \qquad \begin{matrix} 1 \leq i \leq I(k) \\ 1 \leq k \leq P_2 \end{matrix}$$

(5.4) $$\prod_{i=1}^{I(k)} Y_{ik} \leq e, \ Y_{ik} \in \{1, e\} .$$

Since $Y_{ik} \in \{1, e\}$, the problem becomes a combinatorial geometric program. There are two approaches for solving such problems:

(i) Cutting plane via 0–1 mixed linear programming.

(ii) Implicit enumeration.

(i) *0–1 Mixed Linear Programming*

Expansion [11] of a regular geometric program transforms each of the posynomial constraints into a single-term posynomial. If the variables $Z_j = \ln x_j$ and $y_{ik} = \ln Y_{ik}$ replace **x** and **y** the problem becomes an L.P.

Since $Y_{ik} \in \{1, e\}$ the new variable y_{ik} is restricted to the set $\{0, 1\}$. Thus an expansion of E.S.P. (U) transforms the program into a 0–1 mixed L.P. It is possible therefore, to use Gomory's or Bender's algorithms. It should be emphasized that if this approach is adopted two expansions are involved: (1) the original S.P. is expanded into an E.S.P. (U); (2) the E.S.P. (U) is expanded into an L.P. Therefore, each iteration consists of a sequence of 0–1 mixed problems. It is therefore felt that an implicit-enumeration approach will be more appropriate.

(ii) *Implicit Enumeration*

The best method in this approach is still unknown. It is intended, therefore, to give a geometrical interpretation of the method described (sec. 4) and illuminate some of the possibilities through the numerical example.

Every S.P. problem can be transformed into the following form: (see appendix A1):

$$\min x_1 \tag{5.5}$$

$$g_k(\mathbf{x})\, f_k(\mathbf{y}) \leq 1 \qquad k = 1, \ldots, Q_1 \tag{5.6}$$

$$\min_{1 \leq i \leq Q_2} \{h_i(\mathbf{y})\} \geq 1 \tag{5.7}$$

$$(\mathbf{x}, \mathbf{y}) \in (R^m R^{2Q_2})^+ .$$

The functions $g_k(\mathbf{x})$, $f_k(\mathbf{y})$ and $h_i(\mathbf{y})$ are given posynomials and

$$h_i(\mathbf{y}) = y_{2i-1} + y_{2i} \qquad i = 1, \ldots, Q_2 . \tag{5.8}$$

After each iteration

$$h_J(\mathbf{y}^{(k)}) = \min_{1 \leq i \leq Q_2} h_i(\mathbf{y}^{(k)})$$

is expanded around the solution $(\mathbf{y}^{(k)})$:

$$h_J(\mathbf{y}^{(k)}) \Rightarrow \bar{h}_J(\mathbf{y}; \mathbf{U}_J^{(k)}) = \max \left\{ \frac{y_{2J-1}}{U_{2J-1,J}^{(k)}} ; \frac{y_{2J}}{U_{2J,J}^{(k)}} \right\} .$$

The vector $\mathbf{U}_J^{(k)}$ is obtained from (4.1). The new constraint $h_J(\mathbf{y}; \mathbf{U}_J^{(k)}) \geq 1$ is added to the previous constraints.

To avoid cumbersome notation it will be assumed that

$$h_J(\mathbf{y}^{(k)}) = \min_{1 \leq i \leq Q_2} h_i(\mathbf{y}^{(k)}) \qquad \forall k = 1, \ldots, K .$$

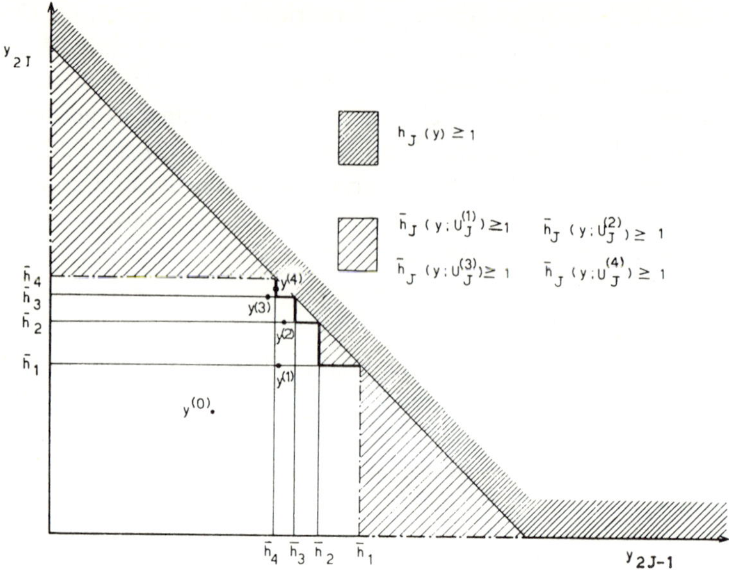

Fig. 1. Schematic representation of the first four iterations: the shaded area is the point set defined by $h_J(\mathbf{y}) \geq 1$; the slanted-line area represents the point set defined by $\bar{h}_J(\mathbf{y}, \mathbf{U}_J^{(i)}) \geq 1$, $i = 1, 2, 3, 4$.

It is seen (fig. 1) that each $h_j(\mathbf{y})$ is approximated by a step-type function. The number of steps in the approximation measures how often the constraint was the most violated one. A step-type point set is described by a conditional constraint. The number of terms in the constraint equals the number of steps plus one: for example (see fig. 1):

After one iteration (one-step approximation)

$$y_{2J-1} \geq u_{2J-1,J}^{(1)} \quad \text{or} \quad y_{2J} \geq u_{2J,J}^{(1)}.$$

After two iterations (two-step approximation)

$$y_{2J-1} \geq u_{2J-1,J}^{(1)} \quad \text{or} \quad y_{2J-1} \geq u_{2J-1,J}^{(2)} \quad \text{and} \quad y_{2J} \geq u_{2J,J}^{(1)} \quad \text{or} \quad y_{2J} \geq u_{2J,J}^{(2)}.$$

As the process evolves either an $\underline{n}$-term conditional constraint becomes an $n+1$-term constraint or a new two-term constraint enters the constraint set. It should be noted that these conditional constraints are lower bounds for the y_1's. As in regular cutting plane procedure a nonactive cut may be discarded; also an $\underline{n}$-term conditional constraint can be reduced to an $(n-k)$-term constraint.

Let F be the following map

$$F: R^{2Q_2} \to \{R, +\infty\}$$

defined by

$$F(\mathbf{U}) = \min\{x_1/(\mathbf{x}, \mathbf{y}) \in (R^m R^{2Q_2})^+, g_k(\mathbf{x}) f_k(\mathbf{y}) \le 1; k = 1, \ldots, Q_1; \quad y_i \ge u_i; i = 1, \ldots, 2Q_2\}.$$

The value of $F(\mathbf{U})$ is $+\infty$ if the corresponding set is empty. Any implicit enumeration is based upon the following property:

$$F(S_1) \le F(S_2) \quad \text{if} \quad S_1 - S_2 = [a_1, a_2, \ldots, a_{2Q_2}] \quad \text{and} \quad a_i \le 0 \quad \forall i.$$

The implicit enumeration is best illustrated through the numerical example.

6. Numerical Example

The example is the famous "Gravel Box" problem [5], [12].

$$\text{(6.1)} \qquad \text{Min } 40 t_1^{-1} t_2^{-1} t_3^{-1} + 20 t_3 t_2 + 40 t_1 t_2 + 10 t_3 t_1$$

subject to

$$\text{(6.2)} \qquad t_1 + t_2 + t_3 \ge 8.$$

The problem was first solved by condensation [1], [12]. It was solved three times starting each time from a different initial point. The results are given in table 6.1. As can be seen the global minimum was obtained at the second

TABLE 6.1

Solutions obtained by the condensation method.
The solution depends on the initial point.

No. Trial	Initial point $t_1^{(0)}$	$t_2^{(0)}$	$t_3^{(0)}$	Solution obtained t_1^*	t_2^*	t_3^*	Value of objective function
I	0.3	8.0	0.5	0.2167	7.3687	0.4155	186.11
II	0.5	0.3	8.0	0.5201	0.2539	7.2260	<u>121.41</u>
III	8.0	0.5	0.3	7.1446	0.1729	0.6835	147.83

trial. The difference between the various solutions is significant. Heuristically it is felt that the implicit enumeration becomes efficient if the difference between the global solution and the local minima is significant.

The implicit enumeration is illustrated in fig. 6.1. The first 16 iterations are shown schematically and numbered. The starting or zero point of the tree represents the solution to the unconstrained problem. The three arcs emanating from this node correspond to the different subproblems solved at the first iteration.

These three subproblems originated from the conditional constraints:

$$t_1 \geq 2.666 \text{ or } t_2 \geq 2.6666 \text{ or } t_3 \geq 2.666 \tag{6.3}$$

obtained by expanding (6.2) with $\mathbf{U}^{(0)} = [1/3, 1/3, 1/3]$. The solution of the first subproblem (minimizing (6.1) subject to $t_1 \geq 2.666$, $(t_2, t_3) \in (R^2)^+$) is 112.78 as can be seen from fig. 6.1. The coordinates of the minimum are $t_1 = 2.666$, $t_2 = 0.308$ and $t_3 = 0.119$. Similarly the solutions to the remaining subproblems were—as can be seen from fig. 6.1—135.357 and 103. The optimal solution for the first iteration is therefore 103. The coordinates of this point are $t_1^{(1)} = 0.7676$, $t_2^{(1)} = 0{,}3299$, $t_3^{(1)} = 2.666$. Since $t_1^{(1)} + t_2^{(2)} + t_3^{(3)} < 8$ the constraint is not satisfied and it was necessary to perform a cut around $\mathbf{t}^{(1)}$. The cut is given by the following constraint:

$$t_1 \geq 1.6314 \quad \text{or} \quad t_2 \geq 0.7113 \quad \text{or} \quad t_3 \geq 5.667. \tag{6.4}$$

The constraint set for the second iteration is given by two conditional constraints ((6.3) and (6.4)). The solution of this problem required solving five subproblems defined by the following constraints:

(i) $t_1 \geq 2.666$	(ii) $t_2 \geq 2.666$	(iii) $t_1 \geq 1.6314 \quad t_3 \geq 2.666$
(iv) $t_2 \geq .7113$	$t_3 \geqslant 2.666$	(v) $t_3 \geq 5.667.$

However, the solution of the first two subproblems is already known. Therefore only three new subproblems have been generated at this iteration—indicated by three arcs leaving node (1). The optimal solution was 106 and it was obtained from subproblem (iv).

The coordinates of the minimum (not included in fig. 6.1) were $t_1^{(3)} = 0.601$, $t_2^{(2)} = 0.711$, $t_3^{(2)} = 2.666$. This point is not feasible, *i.e.*, $t_1^{(2)} + t_2^{(2)} + t_3^{(2)} < 8$. The cut at this iteration is defined by $u_1^{(2)} = 0.151$, $u_2^{(2)} = 0.178$, $u_3^{(2)} = 0.670$ (cf. 4.1) and is given by the following constraint:

$$t_1 \geq 1.208 \quad \text{or} \quad t_2 \geq 1.430 \quad \text{or} \quad t_3 \geq 5.361. \tag{6.5}$$

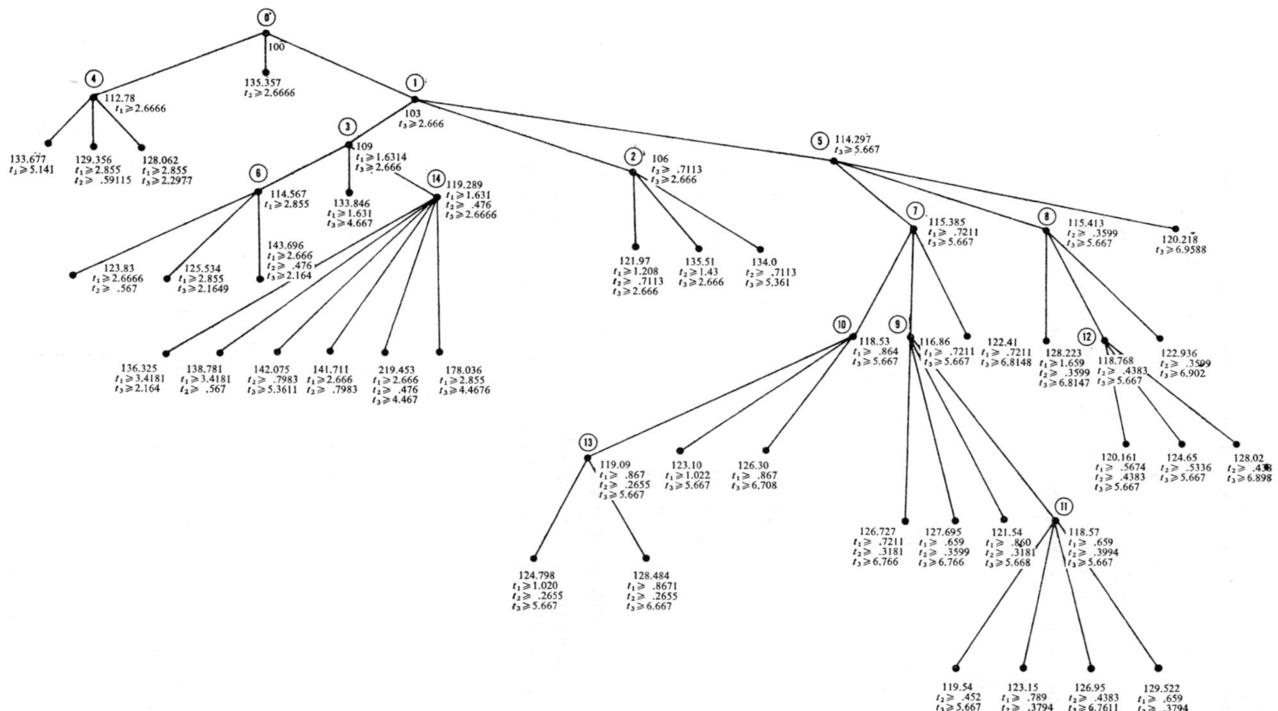

Fig. 6.1. The tree generated after 16 iterations. The numbers in circles indicate iteration numbers. The remaining numbers give the value of the objective function with the associated constraints.

The feasible region for the third iteration is therefore defined by (6.3), (6.4) and (6.5).

The various subproblems solved after each iteration are indicated in fig. 6.1. The value of the optimal solution for the various subproblems is written directly under each node. Under this number are the constraints defining that particular subproblem. As can be seen in fig. 6.1, the average number of subproblems per iteration equals three. Through this explanation, the implicit enumeration used should become clear.

Appendix A1

Consider the following problem.

$$\text{Min } x_1$$

subject to

(A1–1) $$g_k(\mathbf{x}) \leq 1 \qquad k = 1, \ldots, P_1$$

(A1–2) $$h_k(\mathbf{x}) \geq 1 \qquad k = 1, \ldots, P_2$$

(A1–3) $$\mathbf{x} \in (R^m)^+.$$

The first P_1 constraints are called "Regular Constraints", while the second group of P_2 constraints are called "Reversed Constraints".

Both groups of constraints are posynomial functions, and

$$h_k(\mathbf{x}) = \sum_{i=1}^{I(k)} D_{ik} \prod_{j=1}^{m} x_j^{\beta_{ijk}}.$$

Let

$$y_1 = D_{i1} \prod_{j=1}^{m} x_y^{\beta_{iy1}}$$

$$y_2 = \sum_{i=2}^{I(1)} D_{i1} \prod_{j=1}^{m} x_j^{\beta_{iy1}}.$$

If these two variables are introduced, the original constraints (A1–1), (A1–2), and (A1–3) are replaced by

(A1–4) $$\left.\begin{array}{ll} g_k(\mathbf{x}) \leq 1 & k = 1, \ldots, P_1 \\ 1/D_{11} y_1 \prod_{j=1}^{m} x_j^{-\beta_{ij1}} \leq 1 & \end{array}\right\} \text{Regular Constraints}$$

$$\text{(A1.5)}\qquad \begin{cases} h_k(\mathbf{x}) \geq 1 & k = 2, \ldots, P_2 \\ y_2^{-1} \sum_{i=2}^{I(1)} D_{i1} \prod_{j=1}^{m} x_j^{\beta_{ij1}} \geq 1 & \\ y_1 + y_2 \geq 1 . & \end{cases}$$

The addition of the two variables y_1 and y_2 reduced the number of terms in h_1 and added one single-term constraint to the regular constraint group. By repeating this process $2 \sum_{i=1}^{P_2} I(k)$ times one gets the derived form (5.5), (5.7).

Appendix A2

If no Regular Constraints are present, ($P_1 = 0$), the problem becomes a 0–1 mixed linear programming. This follows from (5.3) and (5.4). Let $y_{ik} = \ln Y_{ik}$ and $Z_j = \ln x_j$; then equation (5.3) becomes

$$\sum_{j=1}^{m} \beta_{ijk} Z_j - M_{ik} y_{ik} \leq \ln D_{ik}/u_{ik} \qquad \begin{matrix} 1 \leq i \leq I(k) \\ 1 \leq k \leq P_2 \end{matrix}$$

and equation (5.4) becomes

$$\sum_{i=1}^{I(k)} y_{ik} \leq 1 \qquad y_{ik} \in \{0, 1\} .$$

References

1. Avriel, M. and Williams, A. C. Complementary geometric programming. *S.I.A.M. Journal of Applied Mathematics*, **19**: 125–141, 1970.
2. Carlson, R. C. and Nemhauser, G. L. Scheduling to minimize interaction cost. *Operations Research Journal*, **14**: 52-58, 1966.
3. Cooper, L. Heuristic methods for location allocation problems. *S.I.A.M. Journal of Applied Mathematics*, **6**: 37–53, 1969.
4. Cottle, R. W. and Mylander, W. C. Ritter's cutting plane method for nonconvex quadratic programming. Abadie, J., ed. *Integer and Nonlinear Programming*. North Holland Publishing Cy, Amsterdam, 1970.
5. Duffin, R. J., Peterson, E. L. and Zener, C. *Geometric Programming – Theory and Application*. J. Wiley, New York, 1967.
6. Duffin, R. J. and Peterson, E. L. Geometric programming with signomials. *JOTA* (forthcoming).
7. Falk, J. E. and Soland, R. M. An algorithm for separable nonconvex programming problems. *Management Science*, **15**: 550–569, 1969.

8. Hillier, F. S. and Lieberman, G. J. *Introduction to Operations Research.* Holden-Day, Inc. 1967.
9. Passy, U. Generalization of Geometric Programming; Partial Control of Linear Inventory Systems. *Ph.D. Dissertation.* Stanford University, 1966.
10. Passy, U. and Wilde, D. J. Generalized polynomial optimization. *S.I.A.M. Journal of Applied Mathematics*, **15**: 1344–1356, 1967.
11. Passy, U. Generalized weighted mean programming. *S.I.A.M. Journal of Applied Mathematics*, **20**: 763–778, 1971.
12. Passy, U. Condensing generalized polynomials. *JOTA*, **9**, April 1972.

CHAPTER 5

SENSITIVITY ANALYSIS IN GEOMETRIC PROGRAMMING

M. RIJCKAERT

Summary

The effect of small parameter perturbations on the optimal values of the primal and dual variables can easily be predicted in geometric programming. This chapter will present a new and computationally simple scheme for adjusting the optimal solution vector to small perturbations in the input data of the mathematical model.

The method, which does not depend on how the initial optimal solution is obtained, can be used for both posynomial and signomial problems. A numerical example is presented.

1. Introduction

Geometric programming is a nonlinear programming technique, capable of solving problems that can be formulated as:

$$\text{Minimize } g_0(x)$$

subject to

$$\sigma_m(g_m(x))^{\sigma_m} \leq 1 \qquad m = 1, \ldots, M$$

$$x_n \quad > 0 \qquad n = 1, \ldots, N$$

with

$$g_m(x) = \sum_{t=1}^{T_m} \sigma_{mt} c_{mt} \prod_{n=1}^{N} x_n^{a_{mtn}} \qquad m = 0, 1, \ldots, M.$$

The exponents a_{mtn} (the first index refers to the expression, the second to the term in this expression and the last to a component of x) are arbitrary real numbers and the expressions g_m can always be arranged so that the coefficients c_{mt} are positive. The constants σ_{mt} and σ_m are called signum functions and are restricted to either $+1$ or -1. If all σ_{mt} and σ_m are positive, the problem is termed a posynomial problem. If some σ_{mt} are negative, the problem is said to be in signomial form. Such a problem can be transformed into one with positive σ_{mt} but some negative σ_m, called a reversed geometric program [5].

The difference between posynomial and signomial problems reaches far beyond the nomenclature. Indeed, by simple logarithmic transformation, a posynomial program becomes a convex program. Hence posynomial programs show all the interesting properties associated with convexity. Based on these convexity conditions, a very useful duality theory could be developed for posynomial programs [4].

The presence of some negative signum functions introduces the possibility of obtaining stationary points of various nature and does not guarantee any longer that a local minimum is also global. Last but not least, it also prevents the derivation of a powerful duality theory. A so-called quasiduality theory can be constructed for signomial programs, but it will lack many of the primal-dual properties of posynomials [8]. However, for the present purpose, this quasi-dual will be valuable, the more so since signomial-type problems are more frequently encountered in practice [11]. For simplicity the term dual will be used in both cases and should be seen in the light of the above information.

The mathematical model of a geometric program is mainly determined by two sets of constants: the coefficients c_{mt} on one side and the exponents a_{mtn} on the other side. The exponents usually are derived from physical and economic laws or from experiments relating the different variables of the problem. They are normally well fixed and not subject to change. In contrast, the coefficients c_{mt} are mainly determined from data directly measured or calculated and mostly subject to continuous change.

Hence the following question may be raised: does one need to recalculate the optimal solution upon each variation in c_{mt} or can one get a good approximate solution without repeating the overall computations? Duffin, Peterson and Zener [4] showed that the latter can be done for posynomials. This paper extends their results to signomial programs. The approach is different and an alternative set of sensitivity equations is derived, giving more precise results especially for larger perturbations Δc_{mt} in the values of c_{mt}. A numerical example will be presented to demonstrate the computational efficiency and the economic usefulness of these sensitivity equations.

2. Sensitivity Analysis

To solve the above-stated problem, we assume that the global optimal solution is known for the original value of the coefficients c_{mt}. This solution may have been obtained by any algorithm for solving geometric programs. Among others, algorithms were proposed by Clasen [3] for posynomial programs and for signomials by Avriel and Williams [1], Duffin [6], Hellinckx and Rijckaert [7], Blau and Wilde [2], Passy [10]. Eventually even another nonlinear programming technique may have been used to get the original optimal solution provided that the corresponding dual is known.

A second and important assumption to be made is that the Δc_{mt} are small enough so that the set of active constraints in the optimum will remain unchanged under these perturbations. This implies that Δc_{mt} will not force any previously active constraint to become inactive in the new optimal solution or vice versa.

Finally, we also assume the Δc_{mt} to be such that they won't cause any change in the sign of min $g_0(x)$.

It is obvious that for posynomial problems this third condition might be deleted.

The dual corresponding to the primal global solution has to satisfy the following equations. Note that each dual variable $\omega_{mt}(t=1, \ldots, T_m;$ $m=1, \ldots, M)$ corresponds to one primal term.

1) Normality condition [12]:

$$\sum_{t=1}^{T_0} \sigma_{0t}\omega_{0t} = \sigma_0 . \tag{1}$$

The signum function

$$\sigma_0 = \text{sign}\,(\min g_0(x))$$

is known from the primal solution and is supposed to remain constant under the perturbation Δc_{mt}.

2) N orthogonality conditions [8]:

$$\sum_{m=0}^{M} \sum_{t=1}^{T_m} a_{mtn}\sigma_{mt}\omega_{mt} = 0 \qquad n = 1, \ldots, N . \tag{2}$$

3) D equilibrium conditions [9]:
the number D is called the degree of difficulty of the problem,

$$D = \sum_{m=0}^{M} T_m - N - 1 .$$

The equilibrium conditions are nonlinear equations. They can be stated in logarithmic form as:

$$\text{(3)} \qquad \sum_{m=0}^{M} \sum_{t=1}^{T_m} \sigma_{mt} v_{mt}^d \log\left(\frac{\omega_{mt}}{\omega_{m0}}\right) = \sum_{m=0}^{M} \sum_{t=1}^{T_m} \sigma_{mt} v_{mt}^d \log c_{mt} \qquad d = 1, \ldots, D.$$

The D vectors v^d, $d=1, \ldots, D$ with components v_{mt}^d appearing in these equations are D linearly independent solutions of the homogenized equations (1) and (2) (i.e. obtained by replacing σ_0 in (1) by a zero).

The dual variables ω_{m0} are related to the expressions g_m. They are proportional to the Lagrange multipliers and are called dependent dual variables in contrast to the ω_{mt} which are termed independent dual variables. These dependent dual variables are linear functions of the independent dual variables:

$$\text{(4)} \qquad \omega_{m0} = \sigma_m \sum_{t=1}^{T_m} \sigma_{mt} \omega_{mt} \, .$$

What will happen to the optimum if the coefficients are perturbed by an amount Δc_{mt}? The answer to this question depends on the value of D. If $D=0$, then the optimal dual solution is solely determined by the linear orthogonality and normality equations. These linear equations are not affected by the value of c_{mt} and hence the dual solution remains unchanged. The primal solution however will change and can be derived from

$$\text{(0)} \qquad \begin{aligned} \omega_{mt} &= \frac{c_{mt} \prod_{n=1}^{N} x_n^{a_{mtn}}}{g_0(x)} \qquad && m = 0 \\ \omega_{mt} &= c_{mt} \prod_{n=1}^{N} x_n^{a_{mtn}} && m = 1, \ldots, M. \end{aligned}$$

The primal objective g_0 to be implemented in this formula equals the value of the dual objective and is given by

$$g_0(x) = \sigma_0 \left\{ \prod_{m=0}^{M} \prod_{t=1}^{T_m} \left(\frac{c_{mt} \omega_{m0}}{\omega_{mt}} \right)^{\sigma_{mt} \omega_{mt}} \right\}^{\sigma_0}.$$

If however $D>0$—as in most applications—then the dual solution corresponding to the primal global optimal solution will change too. This is due to the fact that the righthand sides of equations (3) depend on c_{mt}. If c_{mt} changes to $c'_{mt} = c_{mt} + \Delta c_{mt}$, the new dual solution ω'_{mt} will satisfy the following set of equations, analogue to (1–4):

(5) $$\sum_{t=1}^{T_0} \sigma_{0t}\omega'_{0t} = 0$$

(6) $$\sum_{m=0}^{M}\sum_{t=1}^{T_m} a_{mtn}\sigma_{mt}\omega'_{mt} = 0 \qquad n = 1, \ldots, N$$

(7) $$\sum_{m=0}^{M}\sum_{t=1}^{T_m} \sigma_{mt}v^d_{mt}(\log\omega'_{mt} - \log\omega'_{m0}) = \sum_{m=0}^{M}\sum_{t=1}^{T_m} \sigma_{mt}v^d_{mt}\log c'_{mt} \qquad d = 1, \ldots, D$$

(8) $$\sigma_m \sum_{t=1}^{T_m} \sigma_{mt}\omega'_{mt} - \omega'_{m0} = 0 \qquad m = 1, \ldots, M.$$

Subtracting each equation of (1–4) from the corresponding one in (5–8) and using the following Taylor approximation, one gets

$$\log\omega'_{mt} - \log\omega_{mt} \simeq \frac{\Delta\omega_{mt}}{\omega_{mt}},$$

valid for small $\Delta\omega_{mt} = \omega'_{mt} - \omega_{mt}$. After interchanging the sequence of the equations, one gets:

(9) $$\sum_{t=1}^{T_0} \sigma_{0t}\Delta\omega_{0t} = 0$$

(10) $$\sum_{m=0}^{M}\sum_{t=1}^{T_m} a_{mtn}\sigma_{mt}\Delta\omega_{mt} = 0 \qquad n = 1, \ldots, N$$

(11) $$\sigma_m \sum_{t=1}^{T_m} \sigma_{mt}\Delta\omega_{mt} - \Delta\omega_{m0} = 0 \qquad m = 1, \ldots, M$$

(12) $$\sum_{m=0}^{M}\sum_{t=1}^{T_m} \frac{\sigma_{mt}v^d_{mt}}{\omega_{mt}}\Delta\omega_{mt} - \sum_{m=0}^{M}\sum_{t=1}^{T_m} \frac{\sigma_{mt}v^d_{mt}}{\omega_{m0}}\Delta\omega_{m0}$$
$$= \sum_{m=0}^{M}\sum_{t=1}^{T_m} \sigma_{mt}v^d_{mt}(\log c'_{mt} - \log c_{mt}) \qquad d = 1, \ldots, D.$$

For simplicity, denote the dual variables with one index so that ω_{mt} becomes ω_p with

$$p = \sum_{i=0}^{m-1} T_m + t \qquad t \neq 0$$

$$p = \sum_{m=0}^{M} T_m + m \qquad t = 0.$$

The system of equations (9–12) is then schematically represented by

(13) $$A\Delta\omega = b$$

with $\Delta\omega$; b a P-dimensional vector $(P = \sum_{m=0}^{M} T_m + M)$; A a $P \times P$ matrix formed by the coefficients of $\Delta\omega$ in (12).

For nonsingular A,

(14) $$\Delta\omega = A^{-1} b.$$

Since the first $N+1+M$ components of b are identically zero, one only needs to compute the last D columns of A^{-1}, denoted A_D^{-1}, and

(15) $$\Delta\omega = A_D^{-1}\delta$$

with

$$\delta_d = \sum_{m=0}^{M} \sum_{t=1}^{T_m} \sigma_{mt} v_{mt}^d (\log c'_{mt} - \log c_{mt}) \qquad d = 1, \ldots, D.$$

Equations (15) form a first set of sensitivity equations. An alternative one can be obtained by applying the same Taylor approximation to the right-hand side of (13):

(16) $$\Delta\omega = A_D^{-1}\delta'$$

$$\delta'_d = \sum_{m=0}^{M} \sum_{t=1}^{T_m} \sigma_{mt} v_{mt}^d \frac{\Delta c_{mt}}{c_{mt}} \qquad d = 1, \ldots, D.$$

Equations (15) or (16) give an approximate value for the new dual solution. The corresponding primal solution is obtained by neglecting the small gap between the dual and the corresponding primal objective. The above-stated relation (0) between primal and dual variables will produce values for each primal variable. In practice, one shall only compute the independent or control variables this way. The dependent primal variables are then fixed by the mathematical model or by the operation of the system for which the model stands.

In applying the above theory, one should keep in mind that the set of active constraints has to remain unchanged. Therefore one can check, either regularly or else when certain criteria for Δc_{mt} are not met anymore, the inactive constraints. On the other hand, computing ω_{m0} will warn against active constraints becoming loose. Indeed for such constraints the Lagrange multipliers will be zero and hence also ω_{m0}.

3. Numerical Example

The above theory will now be illustrated with the aid of a problem previously solved by Wilde and Passy [9]. From a product *A*, a more valuable product *B* is produced and this product *B* can ultimately react into a waste product *D*:

$$A \rightarrow B \rightarrow D.$$

The reaction rates (i.e. the rates at which product *B* and *D* are formed) are:

$$r_1 = k_1 \exp(-E_1/RT)$$

$$r_2 = k_2 \exp(-E_2/RT).$$

Both reactions occur simultaneously in a series of three perfectly stirred tank reactors of equal volume. The goal is to achieve a maximum yield of product *B* at the end of the third reactor and this for given flow rates of products *A* and *B* at the entrance of the first reactor and a fixed flow rate of *A* at the exit of the third reactor.

It is shown in [9] that this problem can be formulated as:

$$\min\ y_3^{-1}$$

$$\begin{aligned}
\text{s.t.}\quad & x_0^{-1}x_1 + 10x_0^{-1}x_1\theta_1 \le 1\\
& y_0^{-1}y_1z_1 + y_0^{-1}y_1z_1\theta_1^2 \le 1\\
& -(-z_1 - 10y_0^{-1}z_1x_1\theta_1)^{-1} \le 1\\
& -(-x_1^{-1}x_2 - 10x_1^{-1}x_2\theta_2)^{-1} \le 1\\
& y_1^{-1}y_2z_2 + y_1^{-1}y_2z_2\theta_2^2 \le 1\\
& -(-z_2 - 10y_1^{-1}x_2z_2\theta_2)^{-1} \le 1\\
& -(-x_2^{-1}x_3 - 10x_2^{-1}x_3\theta_3)^{-1} \le 1\\
& y_2^{-1}y_3z_3 + y_2^{-1}y_3z_3\theta_3^2 \le 1\\
& -(-z_3 - 10y_2^{-1}x_3z_3\theta_3)^{-1} \le 1
\end{aligned}$$

$$x_1, x_2, x_3, y_1, y_2, y_3, z_1, z_2, z_3, \theta_1, \theta_2, \theta_3 > 0.$$

The variables $x_1, x_2, x_3, y_1, y_2, y_3$ are the flow rates of *A* and *B* after respectively the first, second and third reactor. The variables z_1, z_2, z_3 are auxiliary variables without direct physical meaning and $\theta_1, \theta_2, \theta_3$ are functions of the temperature in the first, second and third reactor. The latter are the independent primal variables and should be manipulated in order to maximize y_3. For the same boundary conditions as in [9],

$$x_0 = 10. \qquad y_0 = 0.1 \qquad x_3 = 0.1,$$

one gets the following dual solution [9]:

$$\begin{array}{lllllll}
\omega_1 = 1.00 & \omega_5 = .099 & \omega_9 = .038 & \omega_{13} = .180 & \omega_{17} = .111 & \omega_{21} = .776 & \omega_{25} = .956 \\
\omega_2 = .149 & \omega_6 = .010 & \omega_{10} = .847 & \omega_{14} = .050 & \omega_{18} = .956 & \omega_{22} = .776 & \omega_{26} = .228 \\
\omega_3 = .569 & \omega_7 = .766 & \omega_{11} = .109 & \omega_{15} = .178 & \omega_{19} = .044 & \omega_{23} = .048 & \omega_{27} = 1.00 \\
\omega_4 = .677 & \omega_8 = .011 & \omega_{12} = .776 & \omega_{16} = .889 & \omega_{20} = .718 & \omega_{24} = .956 & \omega_{28} = 1.00.
\end{array}$$

This corresponds to a primal optimal solution [9]:

$$\begin{array}{lll}
x_1 = 2.077 & y_1 = 7.004 & \theta_1 = 0.382 \\
x_2 = 0.453 & y_2 = 7.646 & \theta_2 = 0.358 \\
x_3 = 0.100 & y_3 = 7.112 & \theta_3 = 0.353.
\end{array}$$

Suppose that the three chemical reactors form just a minor part of a more complex plant and that the product A entering this system of reactors is the outlet of some other production unit. The product B is brought into the process at this stage and its flow rate can therefore be assumed fixed. However it is clear that the value of x_0 might slightly fluctuate, as a result of which also c_{11} and c_{12} will change. This would require a permanent measurement of x_0 and a continuous adaptation of the optimal operating conditions. However because of the above-exposed theory, lengthy computations can be avoided and also a much smaller (and thus cheaper) process computer can be used to adjust the optimal operating conditions. At a given moment a rather steep increase of x_0 from 10.0 to 12.5 occurs. This means that $c'_{11} = 0.08$ instead of $c_{11} = 0.1$ and $c'_{12} = 0.8$ instead of $c_{12} = 1$. The above problem has 7 degrees of difficulty and $P = 19 + 9$. Hence A_D^{-1} is a 28×7 matrix given in appendix.

The vector δ is here:

$$\delta = (0., 0.223, 0.446, -0.223, -0.223, 0., 0.)$$

and

$$\delta' = (0.,\ 0.2,\ 0.4,\ -0.2,\ -0.2,\ 0.,\ 0.).$$

Applying (15) or (16) gives the following numerical results (the bottom figures correspond to formula (16)):

$$\omega_1 = \genfrac{}{}{0pt}{}{1.0}{1.0} \quad \omega_5 = \genfrac{}{}{0pt}{}{.116}{.114} \quad \omega_9 = \genfrac{}{}{0pt}{}{.081}{.077} \quad \omega_{13} = \genfrac{}{}{0pt}{}{.174}{.174} \quad \omega_{17} = \genfrac{}{}{0pt}{}{.130}{.128} \quad \omega_{21} = \genfrac{}{}{0pt}{}{.786}{.785} \quad \omega_{25} = \genfrac{}{}{0pt}{}{.960}{.959}$$

$$\omega_2 = \genfrac{}{}{0pt}{}{.129}{.131} \quad \omega_6 = \genfrac{}{}{0pt}{}{.008}{.008} \quad \omega_{10} = \genfrac{}{}{0pt}{}{.832}{.833} \quad \omega_{14} = \genfrac{}{}{0pt}{}{.057}{.057} \quad \omega_{18} = \genfrac{}{}{0pt}{}{.960}{.959} \quad \omega_{22} = \genfrac{}{}{0pt}{}{.786}{.785} \quad \omega_{26} = \genfrac{}{}{0pt}{}{.277}{.272}$$

$$\omega_3 = \genfrac{}{}{0pt}{}{.547}{.549} \quad \omega_7 = \genfrac{}{}{0pt}{}{.778}{.777} \quad \omega_{11} = \genfrac{}{}{0pt}{}{.127}{.126} \quad \omega_{15} = \genfrac{}{}{0pt}{}{.219}{.215} \quad \omega_{19} = \genfrac{}{}{0pt}{}{.040}{.041} \quad \omega_{23} = \genfrac{}{}{0pt}{}{.103}{.097} \quad \omega_{27} = \genfrac{}{}{0pt}{}{1.0}{1.0}$$

$$\omega_4 = \genfrac{}{}{0pt}{}{.670}{.670} \quad \omega_8 = \genfrac{}{}{0pt}{}{.021}{.020} \quad \omega_{12} = \genfrac{}{}{0pt}{}{.786}{.785} \quad \omega_{16} = \genfrac{}{}{0pt}{}{.870}{.872} \quad \omega_{20} = \genfrac{}{}{0pt}{}{.676}{.680} \quad \omega_{24} = \genfrac{}{}{0pt}{}{.960}{.959} \quad \omega_{28} = \genfrac{}{}{0pt}{}{1.0}{1.0}.$$

The primal independent variables are then

$$\theta_1 = \genfrac{}{}{0pt}{}{0.423}{0.418} \quad \theta_2 = \genfrac{}{}{0pt}{}{0.379}{0.377} \quad \theta_3 = \genfrac{}{}{0pt}{}{0.404}{0.399}.$$

From the mathematical model one gets:

$$x_1 = \genfrac{}{}{0pt}{}{2.390}{2.411} \quad x_2 = \genfrac{}{}{0pt}{}{0.499}{0.505} \quad x_3 = \genfrac{}{}{0pt}{}{0.100}{0.100}$$

$$y_1 = \genfrac{}{}{0pt}{}{8.670}{8.661} \quad y_2 = \genfrac{}{}{0pt}{}{9.260}{9.229} \quad y_3 = \genfrac{}{}{0pt}{}{8.307}{8.303}.$$

The above example illustrates the great computational efficiency of the method. Its usefulness is fully demonstrated by observing the relative values for y_3. In the undisturbed case, y_3 was maximally 7.112. If x_0 changes to 12.5, an optimal value of 7.315 is possible. (This figure is obtained by repeating the optimization procedure for the new data.) Application of formulas (15) or (16) will give values of 8.307 or 8.303. These data show that solutions extremely close to the true optimum could be obtained without repeating the more lengthy and more complex optimization procedure.

Appendix

$$A_{\mathrm{D}}^{-1} = \begin{bmatrix}
0. & 0. & 0. & 0. & 0. & 0. & 0. \\
0.026 & -0.014 & -0.039 & -0.021 & 0.019 & 0. & -0.040 \\
-0.103 & -0.036 & -0.045 & 0.048 & -0.073 & -0.017 & -0.045 \\
-0.087 & -0.037 & -0.012 & 0.111 & -0.137 & -0.028 & -0.010 \\
0.016 & 0. & 0.034 & 0.061 & -0.071 & -0.010 & 0.035 \\
0. & -0.001 & 0. & 0.002 & 0.007 & 0. & 0. \\
-0.071 & -0.036 & 0.022 & 0.170 & -0.215 & -0.038 & 0.025 \\
0. & -0.006 & 0.026 & 0.033 & -0.036 & -0.004 & 0.022 \\
0.007 & 0.020 & 0.080 & 0.110 & -0.125 & -0.017 & 0.087 \\
-0.044 & -0.031 & -0.035 & 0.028 & -0.060 & -0.033 & -0.025 \\
0.034 & 0.026 & 0.023 & -0.034 & 0.022 & -0.011 & 0.038 \\
-0.071 & -0.037 & 0.022 & 0.171 & -0.208 & -0.039 & 0.025 \\
0.061 & 0.033 & -0.034 & -0.177 & 0.170 & -0.006 & -0.012 \\
0.008 & 0.006 & 0.009 & -0.005 & -0.003 & -0.009 & 0.036 \\
0.060 & 0.040 & 0.064 & -0.029 & 0.011 & -0.019 & 0.061 \\
-0.035 & -0.022 & -0.038 & 0.012 & -0.025 & -0.013 & -0.024 \\
0.035 & 0.022 & 0.038 & -0.012 & 0.025 & 0.013 & 0.024 \\
-0.010 & -0.004 & -0.011 & -0.006 & -0.038 & -0.045 & 0.013 \\
0.010 & 0.004 & 0.011 & 0.006 & 0.038 & 0.045 & -0.013 \\
-0.078 & -0.050 & -0.084 & 0.027 & -0.054 & -0.017 & -0.085 \\
-0.071 & -0.037 & 0.022 & 0.171 & -0.208 & -0.039 & 0.025 \\
-0.071 & -0.037 & 0.022 & 0.171 & -0.208 & -0.039 & 0.025 \\
0.007 & 0.014 & 0.106 & 0.143 & -0.161 & -0.021 & 0.109 \\
-0.010 & -0.004 & -0.011 & -0.006 & -0.038 & -0.045 & 0.013 \\
-0.010 & -0.004 & -0.011 & -0.006 & -0.038 & -0.045 & 0.013 \\
0.067 & 0.046 & 0.073 & -0.035 & 0.008 & -0.027 & 0.098 \\
0. & 0. & 0. & 0. & 0. & 0. & 0. \\
0. & 0. & 0. & 0. & 0. & 0. & 0.
\end{bmatrix}$$

References

1. Avriel, M. and Williams, A. C. Complementary geometric programming. *S.I.A.M. Journal of Applied Mathematics*, **19**: 125–141, 1970.
2. Blau, G. and Wilde, D. J. A Lagrangian algorithm for equality constrained generalized polynomial optimization. *AICHE Journal*, **17**: 235, 1971.

3. CLASEN, R. J. The linear-logarithmic programming problem. Rand Corporation, Research Memo, RM-3707-PR, 1963.
4. DUFFIN, R. J., PETERSON, E. L. and ZENER, C. M. *Geometric Programming – Theory and Applications.* J. Wiley & Sons, New York, 1967.
5. DUFFIN, R. J. and PETERSON, E. L. Geometric Programming with Signomials. Carnegie-Mellon University, Report 70-38, 1970.
6. DUFFIN, R. J. Linearizing geometric programs. *S.I.A.M. Review*, **12**: 211, 1970.
7. HELLINCKX, L. J. and RIJCKAERT, M. A solution procedure for geometric programming problems with degrees of difficulty. *Seventh Mathematical Programming Symposium*, The Hague, 1970.
8. PASSY, U. and WILDE, D. J. Generalized polynomial optimization. *S.I.A.M. Journal of Applied Mathematics*, **15**: 1344–1356, 1967.
9. PASSY, U. and WILDE, D. J. Mass action and polynomial optimization. *Journal of Engineering Mathematics*, **3**: 325, 1969.
10. PASSY, U. Finding a global minimum of signomial geometric programming. 1972, this volume.
11. RIJCKAERT, M. Engineering applications of geometric programming. AVRIEL, M., RIJCKAERT, M. and WILDE, D. J., eds. *Perspectives of Optimal Engineering Design.* Prentice Hall, New Jersey, forthcoming.
12. WILDE, D. J. and BEIGHTLER, C. S. *Foundations of Optimization.* Prentice Hall, New Jersey, 1967.

CHAPTER 6

GROUPS, BOUNDS AND CUTS FOR INTEGER PROGRAMMING PROBLEMS

LAURENCE A. WOLSEY

Summary

Various bounds and cuts for integer programming problems have been derived from the asymptotic group problem. Here we point out how a very large range of group problems can be obtained from a given integer problem, allowing one to choose a group problem of arbitrary size to obtain enumerative bounds and a large variety of cuts.

1. Introduction

Various papers have been devoted to obtaining strong bounds and cuts using the group-theoretic structure of the optimal *LP* tableau. In particular, following Gomory's original fractional cutting-plane algorithm [2], and work on the asymptotic problem [3], Glover [1] has shown how to obtain "stronger" cuts; Shapiro [9] has developed enumerative algorithms using group-theoretic bounds; Gorry, Shapiro and Wolsey [6] have shown how relaxation procedures can be used to obtain group problems of manageable size, and very recently Gomory and Johnson [4] have shown how to obtain valid inequalities for general group problems.

Here we indicate how a very wide range of group problems can be obtained from a given integer programming problem, a process which permits us to obtain group problems of manageable and indeed "predetermined" size. These can be used to obtain bounds for use in enumerative algorithms, and also to immediately generate a whole variety of valid cuts [5], and from this "strong" cuts in the sense of [1].

2. Arbitrary Group Problems

Consider the general integer programming (IP) problem:

(P1) $$\min_x \{cx \mid Ax = b,\ x \geq 0 \ \&\ \text{integer}\}$$

where A is $(m \times n)$, b is $(m \times 1)$, x is $(n \times 1)$ and c is $(1 \times n)$. We look first at three possible relaxations of (P1).

Case (*i*) : Let A and b be integer, B the optimal linear programming basis for (P1), $A = (B, N)$, $x = (x_B, x_N)$, $c = (c_B, c_N)$, $\bar{c}_N = c_N - c_B B^{-1} N$, and S an $(m \times r)$ matrix and w an r-vector, then we have a relaxation

$$\text{(P2(i))} \qquad \begin{aligned} &\min c_B B^{-1} b + \bar{c}_N x_N \\ &\text{s.t.} \quad Ax = b + Sw \\ &x \geq 0 \text{ \& integer, } w \text{ integer.} \end{aligned}$$

If $S = B$, problem (P2(i)) is the asymptotic group problem, while if S is a proper submatrix of B we obtain the generalization studied in [10] where nonnegativity is dropped on a subset of the basic variables.

Case (*ii*): As a special case with S square and diagonal we have, taking D_i, $i = 1, 2, \ldots, m$, arbitrary positive real numbers, the relaxed constraints

$$\sum_{j=1}^{n} (a_{ij} \bmod D_i) x_j = (b_i \bmod D_i) \pmod{D_i}, \qquad i = 1, 2, \ldots, m$$

or, dividing through by D_i, and setting $u_{ij} = a_{ij}/D_i - [a_{ij}/D_i]$ and $u_{i0} = b_i/D_i - [b_i/D_i]$,

$$\text{(P2(ii))} \qquad \begin{aligned} &\min \sum_{j=1}^{n} c_j x_j \\ &\sum_{j=1}^{n} u_{ij} x_j = u_{i0} \pmod{1}, \qquad i = 1, 2, \ldots, m, \\ &x_j \geq 0 \text{ \& integer.} \end{aligned}$$

Generalized group problems and the properties of valid cuts have been studied in detail by Gomory and Johnson.

Case (*iii*): Taking the optimal LP tableau for (P1) and dropping the nonnegativity and integrality constraints on all but one of the basic variables x_{Bi}, we obtain the relaxation:

$$\text{(P2(iii))} \qquad \begin{aligned} &\min c_B B^{-1} b + \bar{c}_N x_N \\ &\text{s.t.} \quad x_{Bi} = \bar{a}_{i0} - \sum_{j \in N} \bar{a}_{ij} x_j \\ &x_{Bi}, x_j (j \in N) \geq 0 \text{ \& integer,} \end{aligned}$$

where $\bar{a}_{ij}=(B^{-1}N)_{ij}$, $\bar{a}_{i0}=(B^{-1}b)_i$. The notation $(j\in N)$ indicates that j runs over the index set of the nonbasic variables. We note that Gomory cuts are normally derived from constraints of the above form.

3. Bounding Problems

Henceforth we assume that (P2(iii)), arising from Case(iii) above, is the relaxed problem to be used for generating bounds or cuts. In this section we require that A, b of (P1) are originally integer, or can be made integer by an appropriate transformation. This is always true for problems with inequality constraints, see [6], and holds generally if coefficients are rationals.

The justification and advantage of working with (P2(iii)) to obtain bounds lies in the fact that one constraint of the optimal LP tableau usually determines in large part the integrality of all the basic variables other than x_{Bi}. It is totally accounted for if the underlying asymptotic group problem (AGP) is prime, and often if the AGP is cyclic. (This implies that if (x_{Bi}, x_N) are feasible in (P2(iii)), the corresponding values of the other basic variables $x_B=B^{-1}b-B^{-1}Nx_N$ are integer, and in this case (P2(iii)) is a restriction of the asymptotic group problem.) Let D be the smallest integer such that $D\bar{a}_{i0}$ and $\{D\bar{a}_{ij}\}$ are integer. In the special case cited above we can expect D to equal $|\det B|$, the order of the AGP, while more generally it is a factor of $|\det B|$.

LEMMA 1. For arbitrary real $\lambda>0$,

$$\min \sum_{j\in N} \bar{c}_j x_j+(c_B B^{-1}b)$$

$$\text{s.t.}\quad (D \bmod \lambda)x_{Bi}+\sum (D\bar{a}_{ij} \bmod \lambda)x_j = a_{i0}\ (\text{mod}\ \lambda)\ (\text{mod}\ \lambda)$$

$$x_{Bi},\ x_j(j\in N)\geq 0\ \&\ \text{integer},$$

is a valid relaxation of (P2(iii)), and hence of the original problem (P1).

Setting $\alpha_B=D(\text{mod}\ \lambda)$, $\alpha_j=D\bar{a}_{ij}(\text{mod}\ \lambda)$, and λ a positive integer, a cyclic group problem over the group G_λ of order λ is obtained:

$$\min \sum_{j\in N} \bar{c}_j x_j$$

(P3)
$$\text{s.t.}\quad \alpha_B x_{Bi}+\sum_{j\in N}\alpha_j x_j=\alpha_0\ (\text{mod}\ \lambda)$$

$$x_{Bi},\ x_j(j\in N)\geq 0\ \&\ \text{integer}.$$

This is a shortest-route group problem for which various algorithms exist. Using this approach if the magnitude $|\det B|$ of the asymptotic group problem is too large, a cyclic group problem of manageable and arbitrary size λ can be constructed from which to obtain bounds. In contrast, if $|\det B|$ is small, integrality may give little strength to the bounds, but choosing a larger value of λ is a way of accounting for the nonnegativity of x_{Bi}.

As stated (P3) is likely to have zero-cost solutions if x_{Bi} is unbounded. However, this difficulty is handled by changing the objective function so as still to obtain lower bounds, using elementary duality theory, see [11].

4. Cuts

Choosing an arbitrary λ, we have from the constraint of (P3):

(4) $$\frac{\alpha_B}{\lambda} x_{Bi} + \sum_{j \in N} \frac{\alpha_j}{\lambda} x_j = \frac{\alpha_0}{\lambda} \pmod 1$$

$$x_{Bi},\ x_j \geq 0 \text{ \& integer.}$$

LEMMA 2. Let $\pi_{Bi} x_{Bi} + \sum \pi_j x_j \geq 1$ be any valid inequality for (4), then

(5) $$\sum_{j \in N} (\pi_j - \pi_{Bi} a_{ij}) x_j \geq 1 - \pi_B a_{i0}$$

is a valid inequality in the nonbasic variables.

In particular after first allowing for multiplication of (4) by an arbitrary constant, and taking the valid inequality as $(\alpha_B/\alpha_0) x_{Bi} + \sum (\alpha_j/\alpha_0) x_j \geq 1$, we obtain the fractional cuts of Glover [1].

5. Example

Take as the problem (P2(iii)) (derived from Ex. 3 in [10]):

$$Z = \min \frac{8}{11} x_1 + \frac{9}{11} x_2 + \frac{21}{11} x_3$$

$$\text{s.t.} \quad x_4 = \frac{17}{11} + \frac{4}{11} x_1 - \frac{1}{11} x_2 - \frac{39}{11} x_3$$

$$x_4, x_1, x_2, x_3 \geq 0 \text{ \& integer, } x_4 \leq 3,$$

where $x_{Bi} = x_4$, and $x_N = (x_1, x_2, x_3)$.

Taking $D=11$, the constraint becomes all integer:

$$11x_4 = 17+4x_1-1x_2-39x_3.$$

The AGP is obtained by taking $\lambda=11$, and the resulting solution is $x_3=1$, $Z=21/11$. Arbitrarily taking $\lambda=27$, we obtain the group problem of order 27:

$$\min \frac{8}{11}x_1+\frac{9}{11}x_2+\frac{21}{11}x_3$$

(P3) $$\text{s.t.}\quad 11x_4+23x_1+1x_2+12x_3 = 17 \pmod{27}$$

$$x_4, x_1, x_2, x_3 \geq 0 \text{ \& integer, } x_4 \leq 3,$$

with optimal solution $x_1=4$, $x_4=3$, $Z=32/11$.

To derive a cut we have, after dividing by λ:

(4) $$\frac{11}{27}x_4+\frac{23}{27}x_1+\frac{1}{27}x_2+\frac{12}{27}x_3 = \frac{17}{27} \pmod{1}.$$

Using the procedure for obtaining cuts of [3, 4], we have after normalization the valid cut $\pi_B x_{Bi} + \sum_{j\in N} x_j \geq 1$ of lemma 2:

$$\frac{5}{14}x_4+\frac{4}{7}x_1+\frac{2}{7}x_2+\frac{3}{14}x_3 \geq 1.$$

Finally, after substituting for x_4 we obtain (5), giving after normalization the valid inequality:

$$\frac{36}{23}x_1+\frac{13}{23}x_2-\frac{54}{23}x_3 \geq 1.$$

Although in (P2(iii)) we have only worked with one constraint of the LP tableau, many generalizations are possible. If groups involve more than one basic variable, other bounds and cuts can be obtained, and the result on cuts extends immediately to mixed integer problems. One advantage of working with a single constraint is the simplicity of testing a solution for feasibility in (P2(iii)) before testing in (P1). Additionally, cyclic groups are very easy to handle, and present early experience with an (IP) algorithm based on a group of the form (P3) and a kth best-solution algorithm [11] suggests that an adequate group problem is in practice obtained from one constraint of the LP tableau whenever the original group is too large to handle directly.

References

1. GLOVER, F. Stronger cuts in integer programming. *Operations Research*, **15**: 1174–1177, 1967.
2. GOMORY, R. E. An algorithm for integer solutions to linear programming. Princeton-IBM Mathematics Research Project, Tech. Report No.1, November 1958.
3. GOMORY, R. E. On the relation between integer and non-integer solutions to linear programs. *Proc. Nat. Acad. Sci.*, **53**: 250–265, 1965.
4. GOMORY, R. E. Some polyhedra related to combinatorial problems. *Linear Algebra and Applications*, **2**: 451–558, 1969.
5. GOMORY, R. E. and JOHNSON, E. Some continuous functions related to corner polyhedra. *IBM Research Report* RC 3311, February 1971.
6. GORRY, G. A., SHAPIRO, J. F. and WOLSEY, L. A. Relaxation methods for pure and mixed integer programmming problems. *Management Science*, **18**: 229–239, 1972.
7. HU, T. C. *Integer Programming and Network Flows*. Addison-Wesley, Reading, Massachusetts, 1969.
8. SHAPIRO, J. F. Dynamic programming algorithms for the integer programming problem I: The integer programming problem viewed as a Knapsack type problem. *Operations Research*, **16**: 103–121, 1968.
9. SHAPIRO, J. F. Group theoretic algorithms for the integer programming problem II: Extension to a general algorithm. *Operations Research*, **16**: 928–947, 1968.
10. WOLSEY, L. A. Extensions to the group theoretic approach in integer programming. *Management Science*, **18**, 1971.
11. WOLSEY, L. A. Generalized Dynamic Programming Methods in Integer Programming. Mimeo, CORE, April 1972 (revised January 1973).

CHAPTER 7

A FINITE ALGORITHM FOR HOMOGENEOUS PORTFOLIO PROGRAMMING

JACQUES H. DRÈZE and PAUL VAN MOESEKE[1]

Summary

This article presents a finite algorithm for the computation of an efficient portfolio satisfying the *truncated-minimax rule* (a linear weighting of expectation and standard deviation of yield). This is a problem in homogeneous programming. In fact we solve it by an iteration of parametric quadratic programs: this iteration is kept finite with the help of two simple linear programming procedures, performed after each step.

The *first* procedure is a test indicating that the expected value k of a given parametric quadratic program x^k is greater (say) than the expected value k^* of an optimal portfolio. The *second* procedure computes the smallest (say) expected value of parametric quadratic programs with the base of x^k. A finite number of base shifts then leads to the base of an optimal portfolio, which is found within that base by the calculus.

1. Introduction

We present a finite algorithm for solving the homogeneous portfolio programming problem by an iteration of parametric quadratic programs computed on cuts of the budget set; this iteration is kept finite with the help of two simple linear-programming procedures performed after each step.

The homogeneous portfolio programming problem, introduced by one of us in [7], may be formulated as follows:

$$(H) \qquad \max \phi(x) \equiv cx - (xVx)^{\frac{1}{2}} \text{ on } X \equiv \{x \geqq 0 \mid ix \leq 1\},$$

[1] The authors are grateful to C. Van de Panne for a helpful discussion, out of which some of the ideas in this paper developed; and to S. Gepts, R. Radner and B. von Hohenbalken for their comments.

where c, x, $i \in R^n$; $i \equiv (1, 1, \ldots 1)$, and V is an $n \times n$ positive definite symmetric matrix, so that $\phi(x)$ is concave (see [7], p. 212).

In order to avoid the trivial case where $x = 0$ solves (H), we assume:

$$(P) \qquad \phi(x) > 0 \quad \text{for some} \quad x \in X.$$

A sufficient condition for (P) is that $c_i > (v_{ii})^{\frac{1}{2}}$ for at least one i.

In the portfolio context, c is the vector of expected returns, per dollar of investment, for n securities; V is the covariance matrix of returns, possibly multiplied by a positive scalar reflecting the investor's risk aversion; for further details, see [7].

An x that minimizes xVx subject to $x \in X$ and $cx \geq k$ is usually called "efficient" (see, e.g., Markowitz [3]). We define, for further reference, the quadratic programming problem,

$$(Qk) \qquad \min xVx \text{ on } Xk \equiv \{x \in X \mid cx \geq k\},$$

and denote by x^k a solution to (Qk)—an efficient portfolio with expected return at least equal to k. For the general notion of efficiency see Van Moeseke [8].

As verified below, if x^* solves (H), then x^* also solves (Qk) for $k = cx^*$; and any x^k solving (Qk) for $k = cx^*$ also solves (H). This observation underlies our procedure, which consists essentially in searching for a value k^* such that there exists x^* solving (H) with $cx^* = k^*$, and solving (Qk^*).

Let $K^* \equiv \{k \mid \exists x \text{ solving } (H) \text{ with } cx = k\}$. One readily verifies that K^* is a closed interval[2]. We may thus define $k > (\geq) K^*$ to mean

$$k > \max(\geq \min) \{k \mid k \in K^*\}$$

and $k < (\leq) K^*$ to mean $k < \min(\leq \max) \{k \mid k \in K^*\}$. Our search for an element of K^* is based upon a simple linear programming procedure, introduced in section 2, by which we can test, for an arbitrary k, whether

$$k \begin{Bmatrix} > \\ \in \\ < \end{Bmatrix} K^*.$$

In order to keep our algorithm finite, we rely upon the following property: for x^k solving (Qk), call "base of x^k", and denote Bx^k, the set of

---

[2] Indeed, the set $\{x \in X \mid x \text{ solves } (H)\}$ is nonempty, convex and compact, because X is compact and ϕ is continuous and concave; K^* is the range of a linear function defined on this set.

indices corresponding to the positive coordinates of x^k; call $K(Bx^k)$ the set of values of k for which there exists a solution to (Qk) with the base Bx^k; this set is an (open) interval, the limits of which are easily computed by a simple linear programming procedure introduced in section 3.

This procedure enables us to conduct the search for k^* so as to change the base of x^k at each iteration. Since there exist only finitely many different bases, and no base can reoccur, the algorithm is finite. It terminates either on finding an element $k^* \in K^*$, and giving a solution to (Qk^*); or on finding a base Bx^k such that $K(Bx^k)$ contains K^*. In the latter case, one solves (H) by the calculus procedure presented in [7], p. 249, for maximizing ϕ subject to the equality constraints $ix = 1$ and $x_i = 0$, $i \notin Bx^k$.[3]

After developing the auxiliary linear programming procedures in section 2 and 3, we present our algorithm in section 4. A numerical example is given in section 5.

2. Testing the Optimality of k

We have defined above $Xk \equiv \{x \in X | cx \geq k\}$. We now define similarly $X'k \equiv \{x \in X | cx \leq k\}$, and introduce two auxiliary homogeneous programming problems:

$$(Hk) \qquad \max \phi(x) \text{ on } Xk$$

$$(H'k) \qquad \max \phi(x) \text{ on } X'k$$

as well as their duals, the definition of which involves $\phi_x k$, the gradient of $\phi(x)$ at $x^k (\phi_x k = c - Vx^k \{x^k V x^k\}^{-\frac{1}{2}})$:

$$(Dk) \qquad \min v_1 - kv_2 \text{ on } \{v \geqq 0 | v_1 i - v_2 c \geqq \phi_x k\}$$

$$(D'k) \qquad \min w_1 + kw_2 \text{ on } \{w \geqq 0 | w_1 i + w_2 c \geqq \phi_x k\}.$$

Clearly, for given x^k, (Dk) and $(D'k)$ are simple linear programming problems. We prove in this section the following.

[3] In [7], p. 248, a variant of an infinite algorithm for linear homogeneous programming was applied to problem (H). That algorithm yields a convergent series of portfolios x^t in such a fashion that the set of positive coordinates in x^t is monotone nondecreasing so that coordinates that vanish in x^* become arbitrarily small in x^t as t increases – without necessarily vanishing for finite t, as erroneously stated in the reference. Indeed, there is no telling which coordinates to drop and when, the more so since some coordinates of x^* may be small. See also remark 4A below.

PROPOSITION 2. Let x^k solve (Qk), v^k solve (Dk) and w^k solve $(D'k)$; then

$$v_1^k - kv_2^k \begin{Bmatrix} > \\ = \\ < \end{Bmatrix} w_1^k + kw_2^k \Leftrightarrow k \begin{Bmatrix} < \\ \in \\ > \end{Bmatrix} K^*.$$

The proof of this proposition rests on two lemmas.

LEMMA 2.1. Let x^k solve (Qk). Then

(a) x^k solves (Hk) iff $k \geq K^*$

(b) x^k solves $(H'k)$ iff $k \leq K^*$

(c) x^k solves (H) iff $k \in K^*$.

Proof.

Sufficiency. (a) Let $k \geq K^*$. If x^k does not solve (Hk), then for some $x' \in Xk$, $\phi(x') > \phi(x^k) = cx^k - (x^k V x^k)^{\frac{1}{2}}$. Since x^k minimizes xVx on Xk, this implies $cx' > cx^k \geq k$. Furthermore, for any x^* solving (H),

$$\phi(x^*) \geq \phi(x') > \phi(x^k),$$

and there exists x^* with $cx^* \leq k$ (since $cx^* \in K^*$). Accordingly, there exists $x^0 = \lambda x^* + (1-\lambda)x'$, $0 < \lambda \leq 1$, with $cx^0 = k$, so that $x^0 \in Xk$. By concavity of ϕ, $\phi(x^0) > \phi(x^k)$. This, together with $cx^0 = k \leq cx^k$, implies $x^0 V x^0 < x^k V x^k$, $x^0 \in Xk$, contradicting the fact that x^k solves (Qk). Hence, x^k solves (Hk).

(b) Let $k \leq K^*$. If x^k does not solve $(H'k)$, then for some $x' \in X'k$, i.e. for some x' with $cx' \leq k$, $\phi(x') > \phi(x^k)$. Furthermore, for some x^* with $cx^* \geq k$, $\phi(x^*) \geq \phi(x') > \phi(x^k)$. Accordingly, there exists $x^0 = \lambda x^* + (1-\lambda)x'$, $0 \leq \lambda \leq 1$, with $cx^0 = k$, $\phi(x^0) > \phi(x^k)$, leading to the same contradiction as under (a). Hence, x^k solves $(H'k)$.

(c) Implied by sufficiency of (a) and (b), since $X = Xk \cup X'k$.

Necessity. We first prove (c).

(c) Let x^k solve (H). Then, by definition, $cx^k \in K^*$. We prove that $cx^k = k$. For otherwise, $cx^k > k$, and there exists λ, $0 < \lambda < 1$, with $\lambda x^k \in Xk$ and $(\lambda x^k)V(\lambda x^k) < x^k V x^k$; this contradicts the fact that x^k solves (Qk). Therefore, $k \in K^*$.

(a) Let x^k solve (Hk). If x^k also solves $(H'k)$, then x^k solves (H), and $k \in K^*$, by necessity of (c). If x^k does not solve $(H'k)$, then $k > K^*$, by sufficiency of (b). Therefore, $k \geq K^*$.

(b) The proof is parallel to (a). QED.

LEMMA 2.2. Let x^k solve (Qk), v^k solve (Dk) and w^k solve $(D'k)$; then :

— x^k solves (Hk) iff $\phi(x^k) = v_1^k - kv_2^k$

— x^k solves $(H'k)$ iff $\phi(x^k) = w_1^k + kw_2^k$.

Proof. Immediate consequence of the duality theorem of convex homogeneous programming (theorem 2.3.12 in [7]).

Proof of Proposition 2. Since $x^k \geqq 0$, we may multiply both sides of the constraints in (Dk) by x^k, obtaining:

$$v_1 ix^k - v_2 cx^k = v_2 - kv_2 + [v_1(ix^k - 1) - v_2(cx^k - k)] \geq \phi_x kx^k = \phi(x^k),$$

with the last equality following from homogeneity of degree one of ϕ. This implies, in particular, $v_1^k - kv_2^k \geq \phi(x^k)$; (indeed, the inequalities hold at v^k; and $v_1 - kv_2 \geq v_1 - kv_2 + [v_1(ix^k - 1) - v_2(cx^k - k)]$ since $x^k \in Xk$). By a similar reasoning, $w_1^k + kw_2^k \geq \phi(x^k)$.

We may now combine lemmas 2.1 and 2.2:
by lemma 2.1, x^k must solve either (Hk) or $(H'k)$ or both; and x^k solves (H) iff it solves both (Hk) and $(H'k)$; accordingly, x^k solves (H) iff $\phi(x^k) = v_1^k - kv_2^k = w_1^k + kw_2^k$ and this occurs iff $k \in K^*$; otherwise, either

$$\phi(x^k) = v_1^k - v_2^k < w_1^k + kw_2^k \text{ and } k > K^* \text{ or}$$

$$\phi(x^k) = w_1^k + kw_2^k < v_1^k - kv_2^k \text{ and } k < K^*. \text{ QED.}$$

3. Changing Bases

Let x^k solve (Qk). Bx^k denotes the set of indices corresponding to the positive coordinates of x^k. Let x_+, i_+, c_+ denote the subvectors of x, i, c whose elements have indices in Bx^k; let V^+ denote the matrix formed by the columns of V with indices in Bx^k, and V_+^+ the matrix formed by the rows of V^+ with indices in Bx^k.

Using this notation, we define the following linear programming problem:

(Lk) $\qquad \min k' = c_+ x_+$ subject to:

(3.1) $\qquad V^+ x_+ + \mu i - \nu c \geqq 0$

(3.2) $\qquad V_+^+ x_+ + \mu i_+ - \nu c_+ = 0$

(3.3) $\qquad i_+ x_+ = 1$

(3.4) $\qquad x_+, \mu, \nu \geqq 0.$

It is shown below that (Lk) always has a solution.

We prove in this section the following:

PROPOSITION 3. Let x^k solve (Qk) with $\max_i c_i > k > K^*$ and $\phi(x^k) > 0$, and let k' be given by the solution of (Lk). Then, for every k'', $k \geq k'' > k'$, there exists $x^{k''}$ solving (Qk'') with $Bx^{k''} = Bx^k$; and for every $k'' < k'$, either $k'' < K^*$, or every $x^{k''}$ solving (Qk'') has $Bx^{k''} \neq Bx^k$, or both.

The proof of this proposition rests on the following lemma:

LEMMA 3.1. Let x^k solve (Qk); then:

(a) $cx^k = k$;

(b) $\phi(x^k) > 0$ and $k \geq K^*$ imply $ix^k = 1$.

Proof. (a) This was already proved under lemma 2.1, necessity, (c).

(b) Suppose on the contrary that $ix^k < 1$. Then, for some $\lambda > 1$, $i(\lambda x^k) = 1$ and $(\lambda x^k) \in Xk$ with $\phi(\lambda x^k) = \lambda \phi x^k > \phi(x^k)$. But this contradicts the fact that x^k solves (Hk), as asserted by lemma 2.1 (since $k \geq K^*$). QED.

Proof of Proposition 3. Since Xk has interior points, (Qk) satisfies the Slater condition [5] and x^k solves (Qk) iff it satisfies the Kuhn-Tucker conditions (3.1), (3.2), (3.4) and:

$$i_+ x_+ \leq 1, \ \mu(1 - i_+ x_+) = 0 \tag{3.5}$$

$$c_+ x_+ \geq k, \ v(c_+ x_+ - k) = 0. \tag{3.6}$$

In view of lemma 3.1 (b), x^k satisfies (3.3) as well, so that (3.1)–(3.4) admits at least the solution $x_+^k > 0$. On the compact set $\{x_+ \geqq 0 \mid i_+ x_+ = 1\}$, $c_+ x_+$ reaches a minimum. The linear programming problem (Lk) then has a solution $x_+^{k'} \geqq 0$ with value k'.

Consider any k'' with $k \geq k'' > k'$; then, $k'' = \lambda k + (1-\lambda)k'$ for some λ, $0 < \lambda \leq 1$. (Qk'') has a solution $x^{k''}$ with $Bx^{k''} = Bx^k$. Indeed, define $x_+^{k''} = \lambda x_+^k + (1-\lambda) x_+^{k'}$; $x_+^{k''}$ satisfies (3.1)–(3.4), as a convex combination of solutions to that system, and $c_+ x_+^{k''} = k''$. Accordingly, $x^{k''}$ satisfies the Kuhn-Tucker conditions for (Qk''). Furthermore, $x_+^{k''} > 0$ (since $x_+^k > 0$ and $x_+^{k'} \geqq 0$), so that $Bx^{k''} = Bx^k$.

Consider now any $k'' < k'$. By definition of k', there is no $x_+^{k''}$ solving (3.1)–(3.4) with $c_+ x_+^{k''} = k''$. Accordingly, either (Qk'') does not admit a solution with $Bx^{k''} = Bx^k$; or, if such a solution exists, $i_+ x_+^{k''} < 1$, and $k < K^*$ by (b) of lemma 3.1. QED.

Remarks. 3.A. In general, $k'' \in K(Bx^k)$ for all k'', $k \geq k'' > k'$, but $k' \notin K(Bx^k)$ because $x^{k'} \geqq 0$ but $x^{k'} \ngtr 0$; k' is then a lower limit to $K(Bx^k)$. It may however happen that $x^{k'} > 0$ with k' defining the lower end of $\{k \mid \exists x^k \text{ with } ix^k = 1\}$.

3.B. That $K(Bx^k)$ is an (open) interval is readily verified by considering the two linear programming problems:

$$\max c_+ x_+ \quad \text{and} \quad \min c_+ x_+$$

$$\text{subject to (3.1), (3.2), (3.4) and (3.5)}$$

and reasoning as in the proof of proposition 3, where it is shown that $k \geqq k'' > k'$ implies the existence of $x^{k''}$ with $B(x^{k''}) = B(x^k)$.

3.C. For all $k \geq K^*$, we have seen that x^k solves (Hk). This implies trivially that $\phi(x^{k'}) \geq \phi(x^k)$ for all k, k' such that $k \geq k' \geq K^*$.

4. The Algorithm

In view of proposition 2, we can readily test, for an arbitrary value k, whether $k >, \in$ or $< K^*$; in order to perform the test, we need to solve one quadratic programming problem (Qk) and two linear programming problems, (Dk) and $(D'k)$. Routines for solving quadratic programming problems are well-known and widely available; see e.g., [1] or [6]. Problems (Dk) and $(D'k)$, involving two variables, are almost trivial.

Equipped with this test, one could solve problem (H) quite simply by solving problem (Qk) parametrically in k, generating the whole increasing sequence of values $\{k_1, k_2, \ldots\}$ at which x^k solving (Qk) changes base; for each such k_i, one would test wether $k_i >, \in$ or $< K^*$, until either (i) a $k_i \in K^*$ is found; x^{k_i} then solves (H); or (ii) values k_i, k_{i+1} are found, such that $k_i < K^* < k_{i+1}$. If, for all k with $k_i \leq k < k_{i+1}$, there exists an x^k solving (Qk) with $Bx^k = Bx^{k_i}$, then maximizing $\phi(x)$ subject to the equality constraints $ix = 1$ and $x_j = 0$, $j \notin Bx^{k_i}$ would yield a solution to (H). Given an efficient algorithm for solving (Qk) parametrically in k, this might well be an efficient way of solving problem (H).[4]

[4] T. Hansen has reported orally to one of us that the algorithm he used in solving the quadratic portfolio programming problems for his paper [2] does indeed generate very rapidly the sequence of values $\{k_1, k_2, \ldots\}$ mentioned in the text, and the corresponding vectors x^{k_i}. It thus seems likely that his algorithm could easily be extended from quadratic to homogeneous portfolio programming problems.

In order to avoid generating the whole sequence of values at which x^k changes base, and to permit computations on the basis of any standard quadratic programming routine, we present below a somewhat different algorithm. Essentially, it proceeds from values $\underline{k}$, $\bar{k}$ such that $\underline{k}<K^*<\bar{k}$, and then tests whether $(\underline{k}+\bar{k})/2 >, \in$ or $<K^*$; thereby either terminating, or generating a narrower interval containing K^*. In order to keep the algorithm finite, should K^* consist of a single point, we rely on proposition 3 above: at each step, after finding a new $k \geq K^*$, we compute the lower limit, denoted k', to the set of values k'' for which there exists an $x^{k''}$ solving (Qk'') with $Bx^{k''} = Bx^k$. If $k' \in K^*$, we terminate; if $k'<K^*$, we know that there exists an x^* solving (H) with $Bx^* = Bx^k$; if $k'>K^*$, then k' provides a new $\bar{k}$ and the algorithm continues, but the base Bx^k cannot reoccur, and the finite number of possible bases guarantees finiteness.

Since proposition 3 assumes that k is such that $\max_i c_i > k \geq K^*$ and $\phi(x^k)>0$, an initial step may be necessary to find a k satisfying these two conditions. By remark *3.C*, it then follows that $\phi(x^{k'}) > 0$ for all k' such that $k \geq k' \geq K^*$.

Initial Step

Let $\hat{k} = \max_i c_i$ and test the optimality of $\hat{k}$ (by proposition 2);

(i) if $\hat{k} \in K^*$, set $x^* = x^{\hat{k}}$ and terminate;

(ii) if $\hat{k}>K^*$ and $\phi(x^{\hat{k}})>0$, set $\bar{k} = \hat{k}$, $\underline{k} = 0$ and go to basic step;

(iii) if $\hat{k}>K^*$ and $\phi(x^{\hat{k}}) \leq 0$, let k be the first of the fractions (1/2; 1/4, 3/4; 1/8, 3/8, 5/8, 7/8; 1/16, ...) of $\hat{k}$ such that $k \geq K^*$ and $\phi(x^k) > 0$;[5]
— if $k \in K^*$, set $x^* = x^k$ and terminate;
— if $k>K^*$, set $\bar{k} = k$, $\underline{k} = 0$ and go to basic step.

Basic Step

Let $k = \min_j \{k_j | k_j = \bar{k} - (\bar{k} - \underline{k}) 2^{-j}, k_j \geq K^*, j = 1, 2, \ldots\}$;

(i) if $k \in K^*$, set $x^* = x^k$ and terminate;

---

[5] Condition (P) on p. 80 above guarantees that a satisfactory k can always be found in this way.

(ii) if $k>K^*$, store Bx^k, reset $\underline{k}=2k-\bar{k}$, compute k' by solving (Lk) and test the optimality of k';
— if $k'<K^*$, set $Bx^*=Bx^k$ and go to final step;
— if $k'\in K^*$, set $x^*=x^{k'}$ and terminate;
— if $k'>K^*$, reset $\bar{k}=k'$ and go to basic step.

Final Step

Maximize $\phi(x)$ subject to $ix=1$ and $x_i=0$, $i\notin Bx^*$. Set x^* equal to a solution to this problem and terminate.

THEOREM. The algorithm just described terminates after a finite number of steps, with x^* solving (H).

Proof. (a) *Finite Termination*

The only iteration in the algorithm concerns the basic step and occurs when $k\geq k'>K^*$. Let k_t, k'_t, $\underline{k}_t$, $\bar{k}_t$ denote values generated at the t-th iteration of the basic step (thus, for instance, $\underline{k}_t=2k_t-\bar{k}_{t-1}$). Since $\bar{k}_{t-1}>k_t\geq(\bar{k}_{t-1}+\underline{k}_{t-1})/2>\underline{k}_{t-1}$, we have $\underline{k}_t\geq\underline{k}_{t-1}$, and $K^*>\underline{k}_t$ by definition. If $k'_t>K^*$, then $\bar{k}_{t-1}>\bar{k}_t>\underline{k}_t$ (since $k_t\geq k'_t$). This implies $\bar{k}_t>k_{t+1}$ and, recursively, $\bar{k}_t>k_{t+\theta}$, $\theta=1, 2, \ldots$. Since, by proposition 3, there exists no $k<k'_t$ with $Bx^k=Bx^{k_t}$ and $k\geq K^*$, we know that $Bx^{k_{t+\theta}}\neq Bx^{k_t}$, $\forall\, t, \theta$. Furthermore, the set of bases Bx^k is finite. Accordingly, the sequence $\{\ldots k_t, k_{t+1}\ \ldots\}$ is a decreasing sequence on a finite set, hence a finite sequence, and iteration of the basic step is finite.

(b) x^* *Solves* (H)

x^* is determined in either of two ways:

(i) $x^*=x^k$ for some $k\in K^*$; that x^* solves (H) then follows from (c) of lemma 2.1;

(ii) $k'<K^*<k$ and for all k'', $k'<k''<k$, $x^{k''}$ solving (Qk'') with $Bx^{k''}=Bx^k$; it then follows from (c) of lemma 2.1 and (b) of lemma 3.1 that there is an x^* solving (H) with $Bx^*=Bx^k$ and $ix^*=1$. The calculus procedure of [7], p. 248, can then be applied in the final step to compute such an x^*. QED.

Remarks. 4A. In programming this algorithm for computation one may perhaps find that the calculations are somewhat similar to those suggested by Van de Panne for "Programming with a Quadratic Constraint" [6];

the connection between the two problems seems to be one of duality—see [4]. Combining the suggestion in [4] with the algorithm in [6] might possibly provide a finite algorithm, to be compared with the present one,[6] but this is, as yet, mere conjecture.

4.B. At each step, an upper and a lower bound on $\phi(x^*)$ can be computed as follows: one has $\bar{k} > K^* > \underline{k}$, so that $cx^{\bar{k}} > cx^* = k^* > cx^{\underline{k}}$ for any x^* solving (H); furthermore, $X\underline{k} \supset Xk^* \supset X\bar{k}$ implies $x^{\underline{k}} V x^{\underline{k}} \leq x^* V x^* \leq x^{\bar{k}} V x^{\bar{k}}$; accordingly:

$$\bar{k} - (x^{\underline{k}} V x^{\underline{k}})^{\frac{1}{2}} > k^* - (x^* V x^*)^{\frac{1}{2}} = \phi(x^*) > \underline{k} - (x^{\bar{k}} V x^{\bar{k}})^{\frac{1}{2}}.$$

5. Numerical Example

The following example was computed by hand, using graphical methods to solve the various linear programming problems (in two variables).

Let $c = (1, 2, .5)$, $V = \begin{bmatrix} 1 & 0 & 0 \\ 0 & 2 & 0 \\ 0 & 0 & .5 \end{bmatrix}$, so that

$$\phi(x) = x_1 + 2x_2 - .5x_3 - (x_1^2 + 2x_2^2 + .5x_3^2)^{\frac{1}{2}}.$$

Initial Step

$\hat{k} = \max_i c_i = c_2 = 2$. One solves:

$(Q\hat{k})$: $\min\,(x_1^2 + 2x_2^2 + .5x_3^2)$ subject to $x \geqq 0$, $ix \leq 1$, $x_1 + 2x_2 + .5x_3 \geq 2$. The solution is: $x^{\hat{k}} = (0, 1, 0)$; $\phi(x^{\hat{k}}) = 2 - \sqrt{2}$ and $\phi_x \hat{k} = (1, 2 - \sqrt{2}, .5)$. To test the optimality of $\hat{k}$, one solves:

$$(D\hat{k}) : \min v_1 - 2v_2 \quad \text{s.t.} \quad \begin{aligned} & v_1 - v_2 \geq 1 \\ & v_1 - 2v_2 \geq 2 - \sqrt{2} \\ & v_1 - .5v_2 \geq .5 \\ & v_1, v_2 \geq 0. \end{aligned}$$

The solution is: $v_1 = \sqrt{2}$, $v_2 = \sqrt{2} - 1$, $v_1 - 2v_2 = 2 - \sqrt{2} = \phi(x^{\hat{k}})$.

---

[6] We owe this remark to our colleagues G. de Ghellinck and J. P. Vial.

$$(D'\hat{k}):\ \min w_1+2w_2 \text{ s.t. } w_1+w_2\geq 1$$
$$w_1+2w_2\geq 2-\sqrt{2}$$
$$w_1+.5w_2\geq .5$$
$$w_1, w_2\geq 0.$$

The solution is: $w_1=1$, $w_2=0$, $w_1+2w_2=1>\phi(x^{\hat{k}})$.

By proposition 2, $\hat{k}>K^*$. Since $\phi(x^{\hat{k}})>0$, we set $\bar{k}=2$, $\underline{k}=0$, and go to:

Basic Step

$$k=\min_j \{k_j \mid k_j=2-(2)2^{-j},\ k_j\geq K^*, j=1,2,\ldots\}.$$

— For $j=1$, $k_1=2-1=1$. We compute x^{k_1} by solving:

(Qk_1) $\quad \min(x_1^2+2x_2^2+.5x_3^2)$ s.t. $x\geqq 0$, $ix\leq 1$, $x_1+2x_2+.5x_3\geq 1$.

The solution is: $x^{k_1}=(2/7, 2/7, 2/7)$, $ix^{k_1}=6/7<1$. This reveals immediately that $k_1<K^*$ (see lemma 3.1), so we go on to:

— $j=2$, $k_2=2-.5=1.5$. We compute x^{k_2} by solving:

(Qk_2) $\quad \min(x_1^2+2x_2^2+.5x_3^2)$ s.t. $x\geqq 0$, $ix\leq 1$, $x_1+2x_2-.5x_3\geq 1.5$.

The solution is: $x^{k_2}=(5/13, 7/13, 1/13)$, $\phi(x^{k_2})=1.5-19/\sqrt{494}$, $\phi_x k_2=(1-10/\sqrt{494}, 2-28/\sqrt{494}, .5-1/\sqrt{494})$. To test the optimality of k_2, we solve:

(Dk_2)
$$\min v_1-1.5v_2 \text{ s.t. } v_1-v_2\geq 1-10/\sqrt{494}$$
$$v_1-2v_2\geq 2-28/\sqrt{494}$$
$$v_1-.5v_2\geq .5-1/\sqrt{494},\ v_1, v_2\geq 0.$$

The solution is: $v_1=2-28/\sqrt{494}$, $v_2=0$, $v_1-1.5v_2=2-28/\sqrt{494}>\phi(x^{k_2})$. This reveals immediately that $k_2<K^*$ (see the last statement in the proof of proposition 2), so we go on to:

— $j=3$, $k_3=2-.25=1.75$. We compute x^{k_3} by solving:

(Qk_3) $\quad \min(x_1^2+2x_2^2+.5x_3^2)$ s.t. $x\geqq 0$, $ix\leq 1$, $x_1+2x_2+.5x_3\geq 1.75$.

The solution is: $x^{k_3}=(.25, .75, 0)$, $\phi(x^{k_3})=1.75-\sqrt{19/4}$, $\phi_x k_3=(1-1/\sqrt{19}, 2-6/\sqrt{19}, .5)$. Upon solving (Dk_3) and $(D'k_3)$, we find:

$$v_1=4/\sqrt{19},\ v_2=-1+5/\sqrt{19},\ v_1-1.75v_2=\phi(x^{k_3})$$
$$w_1=1/\sqrt{19},\ w_2=1-2/\sqrt{19},\ w_1+1.75w_2>\phi(x^{k_3}).$$

Accordingly, by proposition 2, $k_3 > K^*$, so that:

$$k = k_3 = 1.75; \; Bx^k = \{1, 2\}; \; \underline{k} = 3.5 - 2 = 1.5 \; (= k_2).$$

To compute k', we solve:

$$(Lk) \qquad \min k' = x_1 + 2x_2 \text{ s.t.} \qquad \begin{aligned} x_1 + \mu - \nu &= 0 \\ 2x_2 + \mu - 2\nu &= 0 \\ \mu - .5 &\geq 0 \\ x_1 + x_2 &= 1 \\ x_1, x_2, \mu, \nu &\geq 0. \end{aligned}$$

The solution is: $x_1 = .4$, $x_2 = .6$, $\mu = .4$, $\nu = .8$, $k' = 1.6$. With $x^{k'} = (.4, .6, 0)$, $\phi(x^{k'}) = 1.6 - \sqrt{22/5}$ and $\phi_x k' = (1 - 2/\sqrt{22}, \; 2 - 6/\sqrt{22}, \; .5)$. To test the optimality of k', we solve (Dk') and $(D'k')$, finding: $v_1 = 2 - 6/\sqrt{22}$, $v_2 = 0$, $v_1 - 1.6v_2 > \phi(x^{k'})$, so that $k' < K^*$. Accordingly, we set $Bx^* = \{1, 2\}$, and go to:

Final Step

Maximize $x_1 + 2x_2 - (x_1^2 + 2x_2^2)^{\frac{1}{2}} - \lambda(x_1 + x_2 - 1)$. To apply formula (3.2.3) of [7], we note that, in the notation of [7]:

$$\bar{c} = (1, 2), \; V = \begin{bmatrix} 1 & 0 \\ 0 & 2 \end{bmatrix}, \; V^{-1} = \begin{bmatrix} 1 & 0 \\ 0 & .5 \end{bmatrix}, \; p = (1, 1), \; m = 1;$$

$$\bar{c}V^{-1}p = 2; \; pV^{-1}p = 1.5; \; \bar{c}V^{-1}\bar{c} = 3;$$

$$\lambda = \frac{2 \pm \sqrt{(2)^2 - 1.5(3-1)}}{1.5} = \frac{2 \pm 1}{1.5} = \text{either } 2 \text{ or } 2/3.$$

By formula (3.2.4) of [7], $x_1 = s(c_1 - \lambda)$, $x_2 = 2s(c_2 - \lambda)$, $s > 0$. With $\lambda = 2$, $x_1 < 0$, $x_2 = 0$. Hence, we choose $\lambda = 2/3$, set $s = 1$ for $x_1 + x_2 = 1$, and get the solution $x^* = (1/3, 2/3, 0)$; $k^* = 5/3$; $\phi(x^*) = 5/3 - 1 = 2/3$.

References

1. Hadley, G. *Nonlinear and Dynamic Programming*. Addison-Wesley, Reading, 1964.
2. Hansen, T. A quarterly portfolio allocation model. Discussion Paper, Series B, Institute of Economics, Bergen, November 1969.
3. Markowitz, H. *Portfolio Selection*. J. Wiley, New York, 1959.

4. SINHA, S. M. A duality theorem for nonlinear programming. *Management Science*, **12**: 385–390, 1965.
5. SLATER, H. Lagrange multipliers revisited: A contribution to nonlinear programming. *Cowles Commission Discussion Paper*, Math. 403, November 1950.
6. VAN DE PANNE, C. Programming with a quadratic constraint. *Management Science*, **12**: 798–815, 1966.
7. VAN MOESEKE, P. Stochastic linear programming: A study in resource allocation under risk. *Yale Economic Essays*, **5**: 196–254, 1965.
8. VAN MOESEKE, P. Towards a theory of efficiency. QUIRK, J. and ZARLEY, A., eds. *Papers in Quantitative Economics*, vol. I. Kansas University Press, 1968.

PART II

SPATIAL PROGRAMMING: NETWORKS

"*Human beings, you know,*" *said Mr. Treves thoughtfully.* "*Human beings. All kinds and sorts and sizes and shapes of 'em. Some with brains and a good many more without. They'd come from all over the place, Lancashire, Scotland—that restaurant proprietor from Italy, and that schoolteacher woman from somewhere out Middle West. All caught up and enmeshed in the thing and finally all brought together in a court of law in London on a grey November day. Each one contributing his little part. The whole thing culminating in a trial for murder.*"

Agatha CHRISTIE. Towards Zero

CHAPTER 8

PLANAR NETWORK-FLOW ALGORITHMS

GUY T. DE GHELLINCK

Introduction

It has long been observed that classical network problems like the max-flow problem can be solved by specific methods when the network is planar. Indeed, most authors point out some specific properties of planar networks, cf. [2], [6], [7], [8], [9]; however, a systematic treatment for a general network-flow problem does not seem to be available in the literature—except in [4].

The purpose of this paper is to emphasize the symmetry between flows and tensions in a planar network and particularly the representation of any flow as a difference of "face numbers". This property will be exploited in conjunction with an algorithm, originally developed by Dijkstra [5], that will be used in both primal and dual form. A class of algorithms is then presented that solve a number of problems such as the max-flow, the transshipment, the min-cost max-flow, and the general piecewise-linear convex problems.

In addition to the complete symmetry of their primal and dual aspects, those algorithms are finite, non-degenerate, and avoid completely the "backtracking" that characterizes other network-flow algorithms.

1. The Network Programming Problem

Consider a network, or finite directed graph consisting of m *nodes* and n *branches*. Assuming that the graph has no loop, i.e. that no branch has the same head and tail, the graph is entirely defined by its $m \times n$ node by branch *incidence matrix* S:

$$(1.1) \qquad S_{ik} = \begin{cases} 1 \text{ if node } i \text{ is head of branch } k \\ -1 \text{ if node } i \text{ is tail of branch } k \\ 0 \text{ otherwise.} \end{cases}$$

The rank of S is $m-1$ if the graph has a single component.

A *flow* is a vector of R^n such that

(1.2) $$Sx = 0.$$

A *tension* y is a vector of R^n such that

(1.3) $$y = uS.$$

When considering each component x_k of x as a quantity flowing in branch k, equation (1.2) is simply a conservation equation for each node.

Considering the vector $u \in R^m$ as a *potential*, equation (1.3) defines the tension on any branch as a difference of potential

(1.4) $$y_{ik} = u_j - u_i.$$

To any branch (i, j) associate further a real convex *function* f_{ij} defined on the closed interval X_{ij}.

The *general network-flow programming problem* is then

(1.5)
$$\min \sum_{ij} f_{ij}(x_{ij})$$
$$\sum_j x_{ij} - \sum_h x_{hi} = 0, \qquad i = 1, 2, \ldots, m$$
$$x_{ij} \in X_{ij} \qquad \text{all } (ij).$$

Since the objective function is convex and the constraints are linear, a necessary and sufficient condition for a vector x^* to be optimal (cf.ch. 1), is that there exists a vector of Lagrange multipliers $u^* \in R^m$ such that the pair (x^*, u^*) is a saddlepoint for the function

(1.6)
$$L(x, u) = \sum_{ij} f_{ij}(x_{ij}) + \sum_{ij} (u_i - u_j) x_{ij};$$
$$L(x^*, u) \underset{u \in R^m}{\leq} L(x^*, u^*) \underset{x \in X}{\leq} L(x, u^*).$$

The condition for a maximum in u is equivalent to constraint (1.5) i.e. requires x^* to be a flow.

The condition for a minimum in x is separable and equivalent to requiring for every branch (ij) that x^*_{ij} solves

(1.7)
$$\min f_{ij}(x_{ij}) + (u_i^* - u_j^*) x_{ij}$$
$$x_{ij} \in X_{ij}.$$

Observe that a convex function g attains its minimum on interval X at x^* if and only if

(1.8) $$g'_-(x^*) \leq 0 \leq g'_+(x^*),$$

where the left (or right) derivative is replaced by $-\infty$ (or $+\infty$) at the lower (or upper) bound of the interval X.

Condition (1.7) reduces then to

(1.9) $$f'_-(x^*_{ij}) \leq u^*_j - u^*_i \leq f'_+(x^*_{ij}).$$

This condition can be given a convenient formulation by introducing the concept of the characteristic of a branch. Consider the correspondence that associates to any $x_{ij} \in X_{ij}$ the interval

$(-\infty, f'_+(x_{ij}))$ if x_{ij} is a lower bound of X_{ij}

$[f'_-(x_{ij}), f'_+(x_{ij})]$ if x_{ij} is interior to X_{ij}

$[f'_-(x_{ij}), \infty]$ if x_{ij} is an upper bound of X_{ij}.

The graph of this correspondence is by definition the *characteristic* C_{ij} of branch (ij) with respect to function f_{ij} and interval X_{ij}.

Using this definition yields:

THEOREM 1.1. A vector x^* solves the general network flow problem (1.5) if:

(i) x^* is a flow

(ii) there exists a tension y^* such that for any branch (ij) the pair $(x^*_{ij}, y^*_{ij}) \in C_{ij}$.

Proof. Straightforward from the preceding argument by writing

$$y^* = u^* S \Leftrightarrow y^*_{ij} = u^*_j - u^*_i .$$

QED.

Observe that the convexity of the function f implies that the characteristic is a monotone curve in R^2. Moreover, the characteristic will be a "staircase" curve iff the function f is piecewise linear (fig. 1.1).

Since any characteristic is monotone, the plane is partitioned into three subsets: the characteristic itself, an upper-left and a lower-right part of the plane.

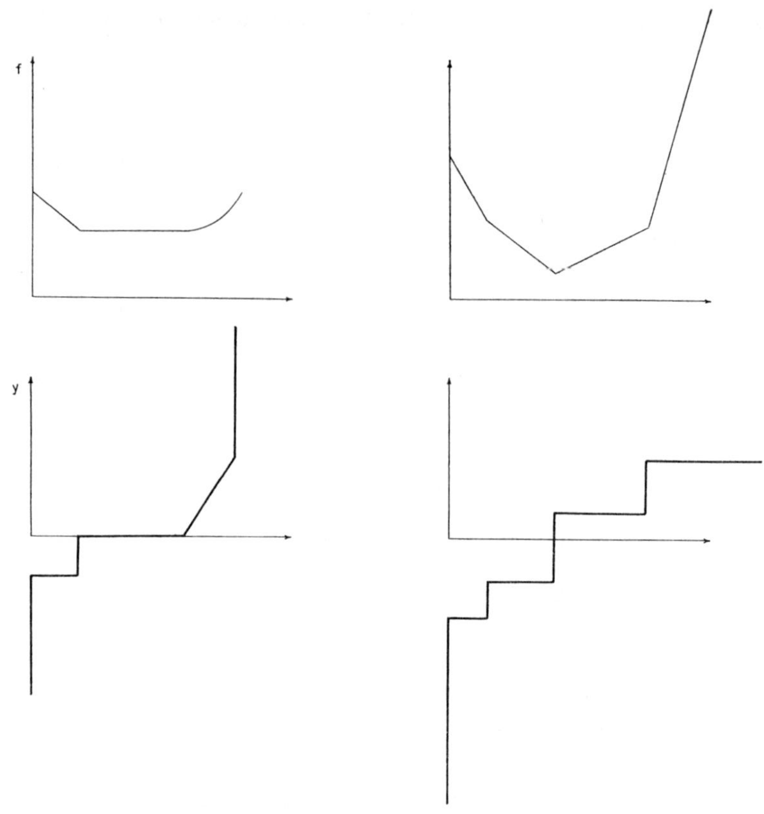

Fig. 1.1

The preceding concepts and terminology can be illustrated by the following example ([2], [8]).

Consider an electrical network consisting of several resistors and, say, one current source and find the current flow consistent with the current source requirement that minimizes the heat or energy losses in this network.

For each resistor with resistance r_{ij} the energy loss function is

$$f_{ij}(x_{ij}) = \tfrac{1}{2} r_{ij} x_{ij}^2; \; x_{ij} \in X_{ij} = (-\infty, \infty).$$

The corresponding (linear) characteristic will be the set of points

$$(x_{ij}, f'_{ij}(x_{ij})) = (x_{ij}, r_{ij} x_{ij}).$$

The current source has a degenerate characteristic consisting of a vertical line.

According to theorem 1.1 the optimality conditions, i.e. the conditions for minimal energy losses are:

(i) $Sx=0 \Leftrightarrow x$ is a flow (Kirchoff's Law)

(ii) for each arc $y_{ij}=u_j-u_i=r_{ij}x_{ij}$ (Ohm's Law).

For a more general analysis of electrical networks see M. Iri [8], sect. 13.

2. Planar Networks

A mathematical characterization of planar graphs has been given by Whitney [12]. An elementary definition can be given as follows [3]. A graph is *planar* when it can be drawn on a plane in such a way that the nodes are distinct points, the branches are simple curves, and no branches intersect except at their endpoints.

The drawing of a planar graph subdivides the plane into *faces*. A connected planar graph with m nodes, n branches has M faces with

$$M=n-m+2.$$

By considering the faces of a planar graph as nodes for another graph and by linking any pair of adjacent faces one defines the *dual graph*. The dual of a planar connected graph is also planar and connected.

Clearly each branch of the primal is also a branch of the dual. The "dual" direction of the branch is defined by considering that it links its right to its left face. (See fig. 2.1.) When a branch links node i to node j and face K to

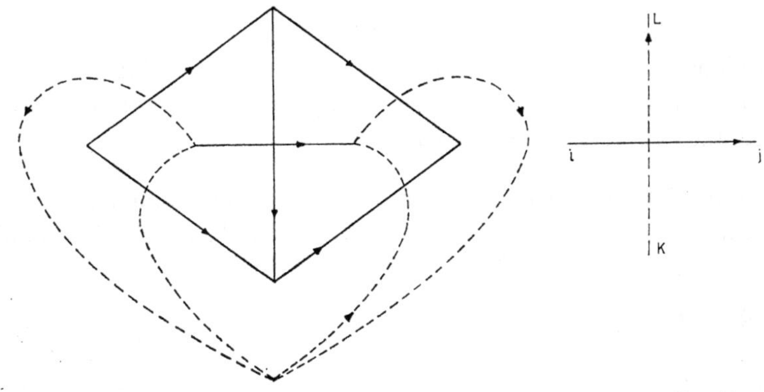

Fig. 2.1

face L either the notation (i, j) or $[KL]$ will be used for referring to this branch.

The dual of the dual graph reproduces the primal with the direction of the branches being reversed.

A fundamental property of a pair of dual planar graphs is the following (for a proof see [3], [8]):

LEMMA 2.1. Let G and G^* be a pair of dual planar graphs. A vector x is a flow on G iff it is a tension on G^*. Conversely a vector y is a tension on G iff it is a flow on G^*.

The correspondence between the basic elements of G and G^* is thus:

Primal G	*Dual G*
Branch	Branch
Node	Face
Face	Node
Flow	Tension
Tension	Flow

Considering that a flow is a dual tension and using the fact that a tension is defined by a potential (see relation (1.4)) one has the particularly simple *representation* of a flow on a planar network.

LEMMA 2.2. A vector $x \in R^n$ is a flow on a planar network iff there is a vector $V \in R^M$ of "face numbers" such that for each branch

$$X_{KL} = V_L - V_K.$$

The optimality condition for the general convex programming problem over a planar network can now be given a simple and symmetric formulation. Indeed, theorem 1.1 can be stated as

THEOREM 2.1. On a planar network a vector x^* is the optimal solution of problem (1.5) iff there exist node numbers u_i, $i = 1, 2 \dots m$ and face numbers V_K, $K = 1, \dots m$ such that for any branch $(ij) = [KL]$:

(i) $x^*_{KL} = V_L - V_K$

$y^*_{ij} = u_j - u_i$;

(ii) the pair (x^*_{ij}, y^*_{ij}) belongs to the characteristic of branch (ij).

Whenever node and face numbers are such that conditions (i) and (ii) hold for branch ij those node and face numbers will be said to be *consistent* with the characteristic of that branch (or "in kilter" according to Ford and Fulkerson [6] terminology).

3. Planar Networks with Step Characteristics

As pointed out in section 1, the characteristic of a branch consists of several steps of either vertical or horizontal intervals whenever the function of that branch is piecewise linear. Considering such a characteristic C it is clear that from any point $(x, y) \in C$ there are always two and only two directions either vertical or horizontal along which the point (x, y) can be moved while remaining on the characteristic. This idea can be used to define nonnegative numbers called "length" for any branch. Such a "primal length" corresponds to a "reduced cost" while a "dual length" corresponds to a "residual capacity".

DEFINITION 3.1. Given a set of node numbers u_i and face numbers V_K consistent with the characteristic of branch $(ij) = [K, L]$ define the primal and dual, forward and reverse lengths by

$$\mathrm{d}_{ij}(u, V) = \sup \{\delta_j \,|\, (V_L - V_K,\, u_j + \delta_j - v_i) \in C_{ij}\}$$

$$d_{ji}(u, V) = \sup \{\delta_i \,|\, (V_L - V_K,\, u_j - u_i - \delta_i) \in C_{ij}\}$$

$$\mathrm{d}_{KL}(u, V) = \sup \{\varepsilon_L \,|\, (V_L + \varepsilon_L - V_K,\, u_j - u_i) \in C_{KL}\}$$

$$d_{LK}(u, V) = \sup \{\varepsilon_K \,|\, (V_L - V_K - \varepsilon_K,\, u_j - u_i) \in C_{KL}\}.$$

Using a symbolism taken from Minty [9], [10] it will be convenient to assign "colors" to the several parts of a stepwise characteristic. (See fig. 3.1.)

Interior of horizontal interval – Red (R).

Interior of vertical interval – Blue (B).

Upper-left corner – Dark Green (DG).

Lower-right corner – Light Green (LG).

Using this symbolism and definition 3.1 for length yields the following table.

TABLE 3.1

	Primal		Dual	
	Forward Δu_j	Reverse Δu_i	Forward ΔV_L	Reverse ΔV_K
Red	0	0	Positive	Positive
Blue	Positive	Positive	0	0
Dark Green	0	Positive	Positive	0
Light Green	Positive	0	0	Positive

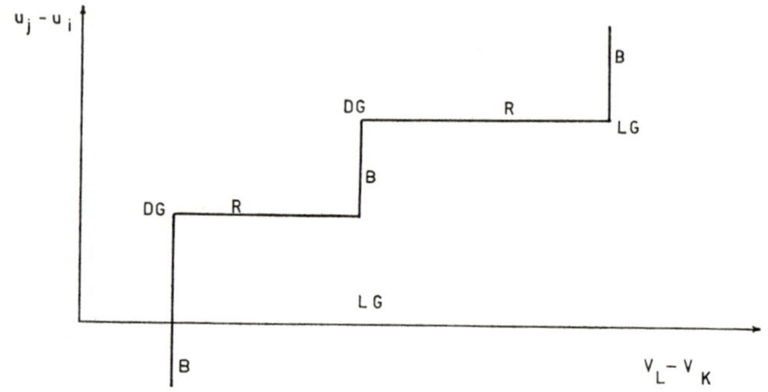

Fig. 3.1

Rephrasing a result of Minty [9], valid for any graph in terms specific to a planar graph, yields:

Lemma 3.1. Consider a planar network whose branches are painted either Blue, Red or Green and let a distinguished branch be painted Green. Then there holds one but not both of the two alternatives:

(i) The distinguished branch belongs to a primal Green and Red cycle, with all Green branches similarly directed.

(ii) The distinguished branch belongs to a dual Green and Blue cycle with all Green branches similarly directed.

The same statement can be given the following extension by further dividing the Green branches into Dark Green and Light Green.

LEMMA 3.2. Consider a planar network whose branches are painted Blue, Red, Light Green or Dark Green and let a distinguished branch be painted Dark Green. Then there holds one but not both of the two alternatives:

(i) The distinguished branch belongs to a primal Green and Red cycle in which all Dark Green branches have a similar direction while all the Light Green branches have the reverse direction.

(ii) The distinguished branch belongs to a dual Green and Blue cycle in which all Dark Green branches have a similar direction while all the Light Green branches have the reverse direction.

Proof. By temporarily reversing the direction of all the Light Green arcs the statement of lemma 2.2 is clearly equivalent to lemma 2.1. QED.

Using those lemmas and the definition 3.1 of lengths yields:

THEOREM 3.1. Given a planar network with step characteristic and a set of node and face numbers consistent with the characteristic of every branch except for a distinguished branch $(m, 1) = [1, M]$ and given the distances $d_{ij}(u, V)$ then there exists one (but not two) of the two alternatives:

(i) There exists a primal path with length zero from node 1 to node m.

(ii) There exists a dual path with length zero from face 1 to face M.

Proof. Paint branch $(m, 1)$ Dark Green and paint all other branches according to the "color" of their characteristics at point $(V_L - V_K, u_j - u_i)$. Clearly, adding branch $(m, 1)$ to the path forms a cycle including the distinguished branch $(m, 1)$ or $[1, M]$. From the length table 3.1, it can then readily be observed that alternatives (i) and (ii) are equivalent respectively to alternative (i) and (ii) of lemma 3.2. QED.

Since the distances are always nonnegative theorem 3.1 can be given the equivalent formulation:

COROLLARY 3.1. Given the assumption of theorem 3.1, there exists one (but not two) of the following:

(i) Every dual path from face 1 to face M has a strictly positive length.

(ii) Every primal path from node 1 to node m has a strictly positive length.

The basic algorithm for planar networks presented in section 6 works by finding the appropriate change of either the face or node numbers that will switch the network from one alternative to the other at each iteration. Those changes result from applying the D algorithm presented in section 4.

4. The *D* Algorithm

In its standard formulation, the shortest-path problem consists in finding a path from node 1 to node m that has the shortest total length. In 1959 Dijkstra [5] proposed an algorithm which is known as the most efficient when all lenghts are nonnegative. A generalization was discussed by Nemhauser [11]. Dijkstra's algorithm works by assigning a label d_j to every node j. When the algorithm is completed d_j represents the length of the shortest path from node 1 to node j. In addition to those labels some "bookkeeping" is needed for backtracking in order to find the actual shortest path.

In this section the emphasis will be exclusively on the labels or node numbers which will rather be considered as the solution of a problem dual to the shortest path.

With the length (definition 3.1) to be used in this paper the case in which no path exists from node 1 to other nodes (i.e. some labels would remain infinite) cannot simply be ruled out. Dijkstra's algorithm has thus been adapted and the rather detailed analysis of this section is necessary to deal with the "partial" solution that corresponds to this case. Consider a network (not necessarily planar) on which nonnegative lengths d_{ij} have been assigned to any branch (or to any pair ij with the understanding that the length d_{ij} is infinite when there is no branch from i to j). Define the set $D \subseteq R^m$ of all vectors d such that

$$\text{(4.1)} \qquad \begin{aligned} & d_1 = 0 \\ & d_k \leq d_h + d_{hk} \qquad h, k = 1, 2, \dots, m. \end{aligned}$$

A vector $d^* \in D$ is *maximal in D* if $d^* \in D$ and $d \in D \Rightarrow d_j \leq d_j^*$ all j.

It will be shown by means of Dijkstra's algorithm that either there exists such a maximal d^* or that the set D is unbounded above. The algorithm will play a basic role for programming problems on a planar network and will be referred to as simply the D algorithm.

D ALGORITHM

Initialization

Set $I^0 = \emptyset$

$$d_j^0 = \begin{pmatrix} 0 : j = 1 \\ \infty : j = 2, \dots, m \end{pmatrix}$$

$p = 1$

Step p

(a) $g^p = \min_{j \notin I^{p-1}} d_j^{p-1}$

if $g^p = \infty$ set

$$\tilde{d}_j = \begin{cases} g^{p-1} & j \notin I^{p-1} \\ d_j^{p-1} & j \in I^{p-1} \end{cases}$$

STOP

otherwise

(b) let $i \notin I^{p-1}$ such that $d_i^{p-1} = g^p$
set $I^p = I^{p-1} \cup \{i\}$

(c) redefine the node numbers by

$d_j^p = \min(d_j^{p-1}, g^p + d_{ij})$

if $p = m$ set $d_j^* = d_j^p$ all j *END*

otherwise go to step $p+1$.

Whenever g^p is finite step p is called "successful"; when g^p is infinite step p will be said to have "failed". In both cases the last vector of node numbers (either d^* or $\tilde{d}$) will be referred to as $\hat{d}$.

LEMMA 4.1. Whenever step p has been completed the following holds:

(i) $d_j^p \leq d_j^{p-1}$ $j = 1, 2, \ldots, m$

(ii) $g^p \leq d_j^p$ $j \notin I^{p-1}$

(iii) $g^p \geq g^{p-1}$

(iv) $d_j^p \leq g^p$ $j \in I^p$

(v) $d_j^p = d_j^{p-1}$ $j \in I^p$.

Proof.

— (i) is obvious from part (c) of step p.

— From (a) of step p: $g^p \leq d_j^{p-1}$, all $j \notin I^{p-1}$; therefore, from (c) and the nonnegativity of d_{ij}

$$g^p \le \min(d_j^{p-1}, g^p + d_{ij}) = d_j^p \text{ all } j \notin I^{p-1}$$

and (ii) holds.

— From (b): $g^p = d_i^{p-1}$ and by (ii): $g^{p-1} \le d_i^{p-1}$ and (iii) holds.

— When $j \in I^p$ one had at some previous step $k \le p$, $d_j^{k-1} = g^k$. By (i) $d_j^k \le d^{k-1} = g^k$ and by (iii) $g^p \ge g^k$; thus (iv) holds.

— By (iv), for all $j \in I^{p-1}$, $d_j^{p-1} \le g^{p-1}$ and by (iii) and the nonnegativity of d_{ij}, $d^{p-1} \le g^p + d_{ij}$; thus by (c) $d_j^p = d_j^{p-1}$, $j \in I^{p-1}$.

Moreover, for $j = i$, $d_i^{p-1} = g^p \le g^p + d_{ii}$ and $d_i^p = d_i^{p-1}$ and (v) holds for all $j \in I^p$. QED.

Remark 4.1. Property (v) of lemma 4.1 actually shows that once node j belongs to the set I^p its node number d^p remains fixed and can thus be viewed as a *permanent label.* As a result, the redefinitions of the labels (step c) should be done only for those $j \notin I^p$.

Another reduction in computational effort can be obtained by separating the nonpermanently labelled nodes ($j \notin I^p$) which have a finite label d_j^p from those that still have their original infinite label and then restricting the search in (a) to the former subset. Also when in (c) a branch ij with zero length is found ($j \notin I^p$) $d_j^p = g^p$ and by (iii) node j will yield a minimum in (a) of the next step, and further search reduction can be achieved.

In order to show inductively that the algorithm will find the maximal element of D when it exists, let us associate to step p the following set $D^p \subseteq R^m$ consisting of all $d \in R^m$ such that

$$(4.2) \qquad d_1 = 0$$

$$d_k \le d_h + d_{hk} \qquad k = 1, 2, \dots, m; \qquad h \in I^p.$$

LEMMA 4.2. After completing part (c) of step p the vector d^p is maximal in D^p.

Proof. After completing step 1 the proposition holds trivially with $I^1 = \{1\}$ and $d_j^1 = d_{1j}$.

Assume inductively that the lemma holds after step $p-1$.

(i) For any $h \in I^{p-1}$ and any k: $d_k^{p-1} \le d_h^{p-1} + d_{hk}$.

Furthermore by lemma 4.1 (i) $d_k^p \le d_k^{p-1}$;

by lemma 4.1 (v) $d_h^p = d_h^{p-1}$, $h \in I^{p-1}$,

and (4.2) holds for d^p and all $h \in I^{p-1}$.

Furthermore by (c) of step p for any k

$$d_k^p \le g^p + d_{ik} = d_i^p + d_{ik}$$

and (4.2) holds for $h = i$ thus for all $h \in I^p$.

This shows that $d \in D^p$.

(ii) Since $I^{p-1} \subseteq I^p$, $D^p \subseteq D^{p-1}$,
consider any $d \in D^p$; by induction $d_j \le d_j^{p-1}$, all j.

If $d_j^p = d_j^{p-1}$ then $d_j \le d_j^p$ (this holds certainly for all $j \in I^p$ by lemma 4.1 (v) and thus also for $j = i$).

If $d_j^p < d_j^{p-1}$ then by part (c) of the algorithm $d_j^p = d_i^p + d_{ij}$ and whenever $d \in D^p$

$$d_j \le d_i + d_{ij} \le d_i^p + d_{ij} = d_j^p.$$

This shows that d is maximal in D^p. QED.

LEMMA 4.3. After completing step p, defining the vector $\tilde{d}$ by:

$$\tilde{d}_j = \begin{cases} d_j^p & j \in I^p \\ K & j \notin I^p \end{cases}$$

with $g^p \le K \le g^{p+1}$, yields

(i) $\tilde{d} \in D$

(ii) $d \in D \Rightarrow d_j \le \tilde{d}_j$, all $j \in I^p$.

Proof.

(i) By lemma 4.2 $d^p \in D^p$ and inequality (4.2) holds for all k and $h \in I^p$. Since $K \le g^{p+1}$, $K \le d_k^p$ (all $k \notin I^p$) and $\tilde{d}_k \le d_k^p$ (all k). Therefore since $\tilde{d}_h = d_h^p$ for $p \in I^p$ one has for $\tilde{d}$:

(4.3) $$\tilde{d}_k \le d_h^p + d_{hk} \text{ all } k \text{ and } h \in I^p.$$

For $h \notin I^p$, when $k \notin I^p$ (4.3) holds trivially from $\tilde{d}_k = \tilde{d}_h = K$; when $k \in I^p$, $\tilde{d}_k = d_k^p \le g^p \le \tilde{d}_h$ and (4.3) holds again. Since (4.3) holds for any h and k, $\tilde{d} \in D$.

(ii) Since any $d \in D$ also belongs to D^p and by lemma 4.2 d^p is maximal in D^p:

$$d \in D \Rightarrow d_j \leq d_j^p = \tilde{d}_j \quad \text{all } j \in I^p. \text{ QED.}$$

Since in case of "failure" arbitrarily large values of K can be selected while defining a vector $d \in D$, lemmas 4.2 and 4.3 prove the following.

THEOREM 4.1. Let $\hat{d}$ be the vector generated by the D algorithm, i.e. either $\hat{d} = d^*$ (success) or $\hat{d} = \tilde{d}$ (failure at step $p+1$). Then $\hat{d}$ belongs to D and

(i) if $\hat{d} = d^*$ the set D has a unique maximal element $\hat{d}$;

(ii) if $\hat{d} = \tilde{d}$ the set D is unbounded and $\hat{d}$ is maximal with respect to its components $i \in I^p$.

The node numbers $\hat{d}$ generated by the D algorithm can also be used to define a new set of "reduced lengths" in the following sense.

COROLLARY 4.1. Define the reduced length $\bar{d}_{hk}$ by

$$\bar{d}_{hk} = d_{hk} + \hat{d}_h - \hat{d}_k \tag{4.4}$$

and the corresponding set $\bar{D}$ as defined by equation (4.1) with length $\bar{d}$. Then:

(i) $\bar{d}_{hk} \geq 0$ all h, k

(ii) the vector $d = 0$ belongs to $\bar{D}$

(iii) $d \in \bar{D} \Rightarrow d_j \leq 0$ all j (when $\hat{d} = d^*$)
$d_j \leq 0$ $j \in I^p$ (when $\hat{d} = \tilde{d}$).

Proof. Combining the definition (4.4) of $\bar{d}$ with the inequality (4.1) shows both that $\bar{d}_{hk} \geq 0$ and

$$d \in \bar{D} \Leftrightarrow (d_k + \hat{d}_k) \leq (d_h + \hat{d}_h) + d_{hk} \quad \text{all } h, k.$$

Therefore $d \in \bar{D} \Leftrightarrow (d + \hat{d}) \in D$.

The maximality of d^* in D (lemma 4.2) or partial maximality of $\tilde{d}$ in D (lemma 4.3) are thus equivalent to the same property for $d = 0$ in $\bar{D}$. QED.

Comparing the previous results with the more usual "primal" interpretation of the algorithm [5], [11] it can readily be observed that the quantities d^* generated by the algorithm represent the shortest distances from node 1 to node j. When the algorithm "fails" the nodes $j \in I^p$ are those that cannot be reached from node 1 through a path with finite length.

On the other hand the definition of the "reduced length" can be seen as a method for generating nonnegative lengths such that there is a path with zero length connecting node 1 to any node j (success) or at least to any node $j \in I^p$ (failure).

5. The Max-Flow Problem on a Planar Network

The max-flow problem is usually defined in the following sense. Consider a network in which each branch has a (possibly infinite) capacity $C_{ij} \geq 0$; the problem is to force the maximum amount from a source to a sink without exceeding the branch capacities. Observe that the conservation equation at the source and sink nodes imposes that a "special branch" going from sink to source should be added to the network. The problem is then to find a flow (as defined by 1.2) whose component on the special branch is maximum while capacity constraints are met.

The purpose of this section is to show how this problem can be solved by a single application of the D algorithm whenever the network including the special arc is planar. Since the algorithm will operate on the dual graph (even though only its primal representation needs to be used) branches will be referred to by the faces they link.

Index the faces in such way that the special branch becomes branch $[1, M]$ with $C_{1M} = \infty$. To any branch $[KL]$ assign the nonnegative lengths

$$d_{KL} = C_{KL} \tag{5.1}$$

$$d_{LK} = \begin{cases} 0 & \text{for a directed branch} \\ C_{KL} & \text{for an undirected branch.} \end{cases} \tag{5.2}$$

THEOREM 5.1. Applying the D algorithm to the dual network with lengths defined by (5.1), (5.2) yields either:

(i) a vector $d^* \in R^m$ (success) and the flow is optimal:

$$x^*_{KL} = d^*_L - d^*_K.$$

(ii) a vector $\tilde{d} \in R^m$ (failure) and a set I^p:

(a) if $M \in I^p$ the vector $\tilde{x}_{KL} = \tilde{d}_L - \tilde{d}_K$ is optimal

(b) if $M \notin I^p$ the flow on $[1, M]$ is unbounded.

Proof. By lemma 2.2 any flow on the planar network can be defined as a difference of face numbers d_K.

The flow to be maximized is $x_{1m}=d_M-d_1$.

The capacity constraints become:

$$x_{KL}\leq C_{KL}\Leftrightarrow d_L-d_K\leq C_{KL}. \tag{5.3}$$

$$x_{KL}\geq\begin{cases} 0 & \Leftrightarrow d_K-d_L\leq 0\\ -C_{KL} & \Leftrightarrow d_K-d_L\leq C_{KL}; \end{cases} \tag{5.4}$$

and by using the definition (5.1), (5.2) of length the capacity constraints (5.3), (5.4) have exactly the form (4.1) used to define the set D.

Therefore, any $d\in D$ yields a feasible flow and a maximal $d\in D$ will define a maximum flow as far as it maximizes the component d_M of d (since $d_1=0$ and $x_{1M}=d_M-d_1$). QED.

Observe that this method for obtaining a max-flow solution contrasts with the Ford-Fulkerson [6] method for a general network and with the Berge [2] method for a planar network in this respect:

— The method requires at most M steps and convergence is thus ascertained without requiring the capacities to be rational numbers.

— The actual maximal flow is derived without requiring any backtracking nor adding up flows on flow-augmenting paths.

Observe also that the method handles the case of a directed or undirected network as well.

Figure 5.1 shows an example for an undirected network. Numbers in brackets are the branch capacities. Each face contains its sequence of decreasing face numbers; the starred face numbers define the max-flow.

6. A Basic Algorithm for Planar Networks

The algorithm presented in this section will solve a problem of the form (1.5) on a planar graph with piecewise linear convex functions (monotone step-characteristics) when the following initial conditions hold:

— Indices given to nodes and faces are such that a special branch is labelled $(m, 1)$ and $[1, M]$.

— Node and face numbers u_j and V_K are available which are consistent with the characteristics of all but the special branch

$$(V_L-V_K,\, u_j-u_i)\in C_{ij}\quad (\text{all } (ij)=[KL]\neq(m,1)). \tag{6.1}$$

MAX-FLOW ALGORITHM
1
[12]
1 SINK
∞
45*
[12]
∞
40*
[40]
[18]
∞
27*
∞
36*
[13]
SOURCE
[15]
1
[30]
1
∞
13
12*
∞
6*
[6]
[13]
[6]
0
0*
[∞]

ACTUAL FLOWS
45 - 40 = 5
45
45-27 = 18
36 - 27 = 9
40 - 36 = 4
40 - 0=40
45
27-12 = 15
36 - 6 = 30
12 - 0 = 12
12 - 6 = 6
6 - 0 = 6

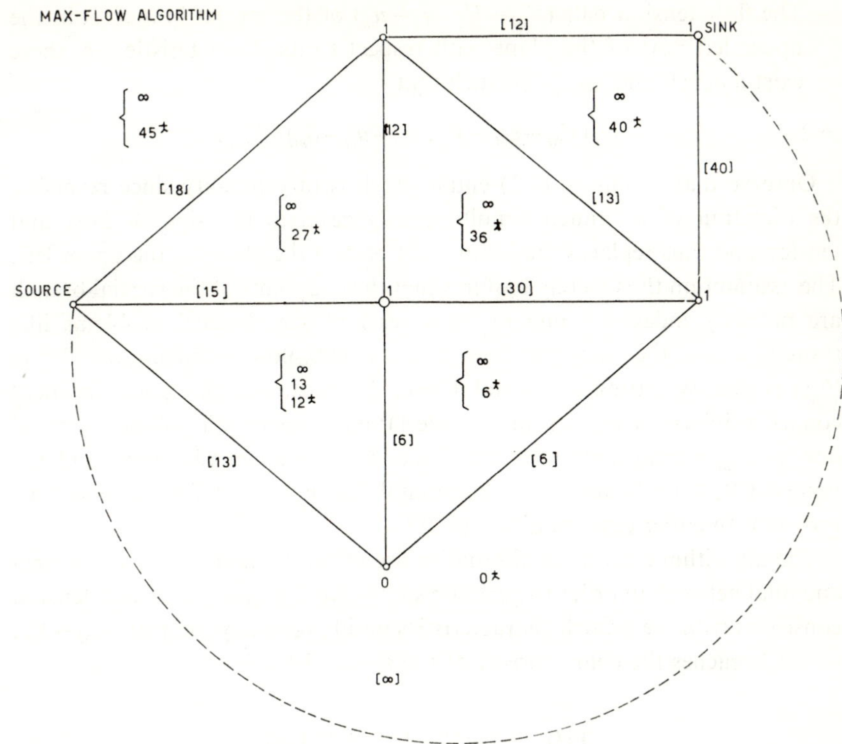

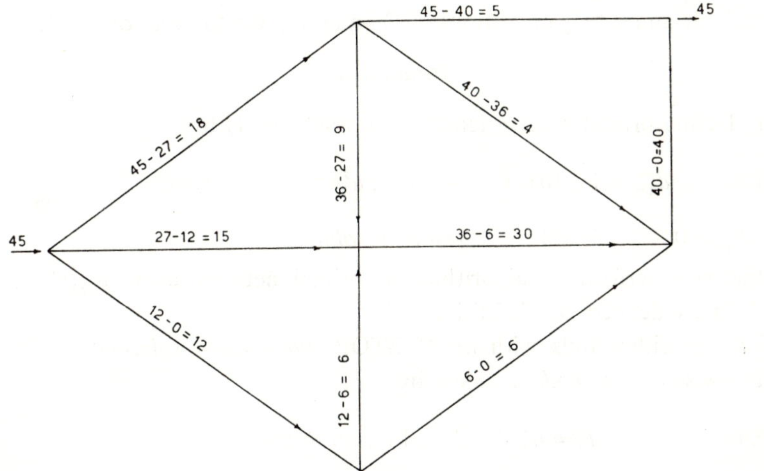

Fig. 5.1

— The flow tension pair $(V_M - V_1, u_1 - u_m)$ of the special branch is in the upper left part of the plane with respect to its characteristic, i.e. there exist a $\delta_m \geq 0$ and an $\varepsilon_M \geq 0$ such that

(6.2) $$(V_M + \varepsilon_M - V_1, u_1 - u_m - \delta_m) \in C_{m1}.$$

Observe that condition (6.2) entails no loss of generality since reversing the direction of a branch simultaneously reverses the sign of flow and tension and thus replaces the lower right part of the plane by the upper left. The assumption thus merely requires that the endpoints of the special branch are properly indexed 1 and m. In several of the classical problems like transshipment, the min-cost flow and the max-flow problems, condition (6.1) is met by setting $u = 0$ and $V = 0$. A return branch (special branch) connects sink (node m) to source (node 1); its characteristic will be a vertical line at x_{1M} = total supply = total demand for the transshipment and the min-cost flow problems and a horizontal line $y_{m1} = -1$ for the max-flow problem. In either case condition (6.2) holds.

The algorithm consists of alternative use of the D algorithm to the primal and dual network in order to generate successive changes of flow and tension consistent with the branch characteristics untill eventually the pair $(V_M - V_1, u_1 - u_m)$ reaches the characteristic of the special branch.

THE BASIC ALGORITHM

Let u_j° and V_K° be the node and face numbers which are initially available and satisfy conditions (6.1) and (6.2). Go to iteration $s = 1$.

Iteration s

(a) Define primal reverse length of branch $(m, 1)$ by

(6.3) $$d_{1m} = \inf \{\delta \mid (V_M^{s-1} - V_1^{s-1}, u_1^{s-1} - u_m^{s-1} - \delta) \in C_{m1}\}$$

if $d_{1m} = 0$ END (*optimal solution*)
otherwise perform D algorithm on primal network using $d_{ij}(u^{s-1}, V^{s-1})$ as defined by definition 3.1
if D algorithm fails with $m \notin I^p$ STOP (*no feasible solution*)
otherwise *change node numbers* by

(6.4) $$u_j^s = u_j^{s-1} + \hat{d}_j \text{ (all } j); \text{ go to } (b).$$

(b) Define dual forward length of branch [1, M] by

(6.3)′ $$d_{1M} = \inf\{\varepsilon \mid (V_M^{s-1} + \varepsilon - V_1^{s-1}, u_1^s - u_m^s) \in C_{1M}\}$$

if $d_{1M} = 0$ END (*optimal solution*)

otherwise perform D algorithm on dual network using $d_{KL}(u^s, V^{s-1})$ as defined by definition 3.1

if D algorithm fails with $M \notin I^p$ STOP (*unbounded objective function*)

otherwise *change face numbers* by

(6.4)′ $$V_K^s = V_K^{s-1} + \hat{d}_K \text{ (all } K)$$

go to iteration $s+1$.

Lemma 6.1. After every change of node (or face) numbers as defined by (6.4) (or (6.4)′):

(i) conditions (6.1) and (6.2) hold

(ii) there exists a path from node 1 to node m (or from face 1 to face M) that has length zero.

Proof. Consider the case of "primal" step (a), i.e. a change of node numbers.

(i) By theorem 4.1 the D algorithm generates a vector $\hat{d} \in D$; this implies for all branches (i, j)

$$-d_{ji}(u^{s-1}, V^{s-1}) \leq \hat{d}_j - \hat{d}_i \leq d_{ij}(u^{s-1}, V^{s-1}).$$

By definition 3.1 of a length

$$(V_L^{s-1} - V_K^{s-1}, u_j^{s-1} - u_i^{s-1} + \hat{d}_j - \hat{d}_i) \in C_{ij}$$

and since $u^s = u^{s-1} + \hat{d}$ condition (6.1) holds for V^{s-1} and u^s. Moreover $\hat{d} \in D$ implies $\hat{d}_m \leq d_{1m}$ and by definition (6.3) of d_{1m} condition (6.2) that was assumed inductively to hold after step s will hold after step $s+1$.

(ii) By definition 3.1 of a length observe that

$$d_{ij}(u + \hat{d}, V) = d_{ij}(u, V) + \hat{d}_i - \hat{d}_j.$$

The new length is thus the reduced length of corollary 4.1. This shows the existence of a path of length zero from 1 to m. Exactly the same argument applies for "dual" step (b). QED.

THEOREM 6.1. Each step of the basic algorithm (except perhaps the very first) yields a strictly positive increase of either u_m or V_M. After finitely many steps the algorithm will terminate at one of the three possibilities:

(i) END (a) or (b): an optimal solution has been found;

(ii) STOP (a): the problem has no feasible solution, i.e. there exists no flow x with $x_{ij} \in X_{ij}$ (all ij);

(iii) STOP (b): there exist feasible solutions but the objective function has no lower bound.

Proof. By lemma 6.1, after completing step (a) (or (b)) there is a path of zero length from node 1 to node m (from face 1 to face M) and by theorem 3.1 any dual (primal) has now positive length. Therefore any step (a) (or (b)) is such that the following step (b) (or (a)) will strictly reduce the length of a dual (primal) path, i.e. will yield a strictly positive increase of V_M (or u_m) since u_1 and V_1 will never change. Thus cycling cannot occur. Moreover the existence of a path of length zero implies that for each branch of that path the difference $u_j - u_i$ (or $V_L - V_K$) is exactly equal to the level of a horizontal (or vertical) interval of its characteristic. Only finitely many such values exist and since $u_m - u_1$ $(V_M - V_1)$ is the sum of less than $m(M)$ such differences and since u_1 and V_1 are held fixed, u_m (or V_M) can be given only a finite number of values and the number of iterations is finite.

(i) When END occurs at step (a) or (b) an optimal solution has been found since the node and face numbers are now consistent with the characteristics of all branches including the special branch $(m, 1)$.

(ii) When the algorithm reaches the STOP in step (a) any primal path from node 1 to node m has infinite length. Considering the definition of the length, any path from node 1 to node m contains at least one branch (ij) which is either positively directed with flow at the upper bound of X_{ij} or reversely directed with flow at its lower bound. No further increase of flow from m to 1 is feasible. Moreover length d_{1m} itself is infinite and the present flow from node m to node 1 is still below its required lower bound.

(iii) When the algorithm reaches the STOP in step (b) an arbitrarily large increase $\tilde{\delta}(K)$ can be given to face number V_M while maintaining consistency condition (6.1) by lemma 4.3. The flow $x_{m1} = V_M^{s-1} + \tilde{\delta}_M - V_1^{s-1}$ increases without bound with K. The effect of such a change on each function

$$f_{ij}(x_{ij}) + (u_i^s - u_j^s)x_{ij}$$

is zero for ordinary branches since (6.1) implies $f'_{ij}(x_{ij})+u_i-u_j=0$. For the special branch, $d_{1m}=\infty$ implies by (6.3)′ that

$$f'_{m1}(x_{m1})+u^s_m-u^s_1<0;$$

the increase of K and thus x_{m1} results in an unbounded decrease of the Lagrangean function $L(x, u)$ (see (1.6)). But whenever the vector x is a flow the value of the Lagrangean is the value of the objective function itself. QED.

The Simple Transshipment Problem

The problem consists in shipping from supply nodes with available supplies d_i to meet demand d_j at demand nodes. Let D equal total demand equals total supply. Let c_{ij} represent a nonnegative unit cost on any "real" branch and connect any supply node i to a joint source node, say 1, through an auxiliary branch $(1, i)$ and any demand node j to a joint sink through an auxiliary branch (j, m). Finally connect the sink to the source by a special branch $(m, 1)$. Assume that this *extended network* is planar.

The characteristics of the three kinds of branches are as follows (fig. 6.1). Since both kinds of characteristics contain the point (0, 0) the algorithm can start with $u_j=0; j=1, \ldots m$; $V_K=0$; $K=1, 2 \ldots M$.

The General Min–Cost Flow Problem [6], [4]

This problem is basically the same as the transshipment problem except that the "real" branches might also have a limited capacity, say b_{ij}, or more generally a piecewise linear convex cost structure.

In such case the characteristic contains two or more vertical parts while still containing the point (0, 0) and the same initialization can be used to start the basic algorithm.

A discussion of the use of the basic algorithm in this problem with the description of a computer code and numerical experience is presented in [4].

The Max–Flow Min–Cost Problem [6]

In this problem capacity restrictions are imposed upon some of the arcs while supply and demand requirements are not fixed but rather represent upper bounds to either supply or demand. The problem is then to ship the maximum quantity through the network at minimum cost. With this for-

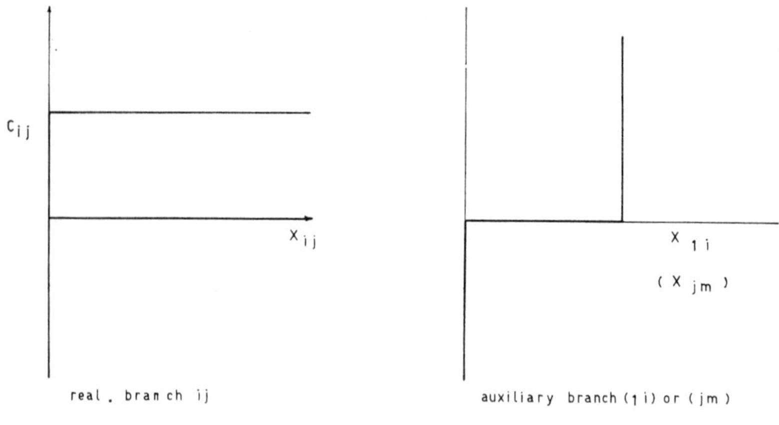

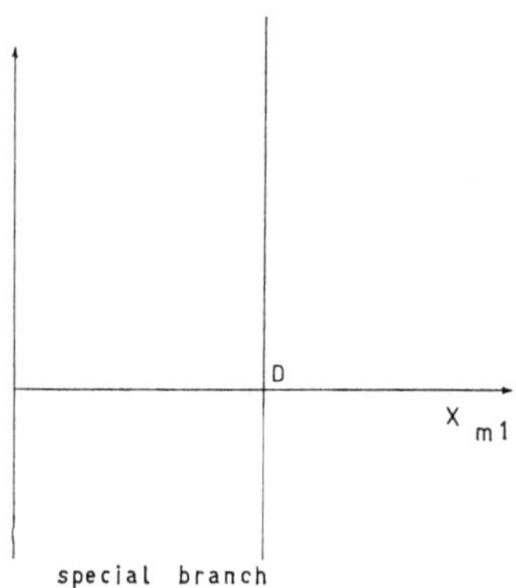

Fig. 6.1

mulation the characteristic of the special branch is not defined and can be thought of as lying at infinity on the "right" and at the "bottom" of the plane in such way that definition (6.3) of d_{1m} and (6.3)′ of d_{1M} will always yield $+\infty$. When the algorithm stops in (a) the solution at hand provides the maximum flow through the network at minimum cost. When the algorithm stops in (b) the flow is unbounded: any flow above V_M^{s-1} can be achieved by using $V^s = V^{s-1} + \tilde{d}(K)$ for arbitrarily large K and any such flow is achieved at minimum cost.

Figure 6.2 gives an example of a max-flow min-cost problem; numbers in brackets are the branch capacities while the numbers in parentheses are

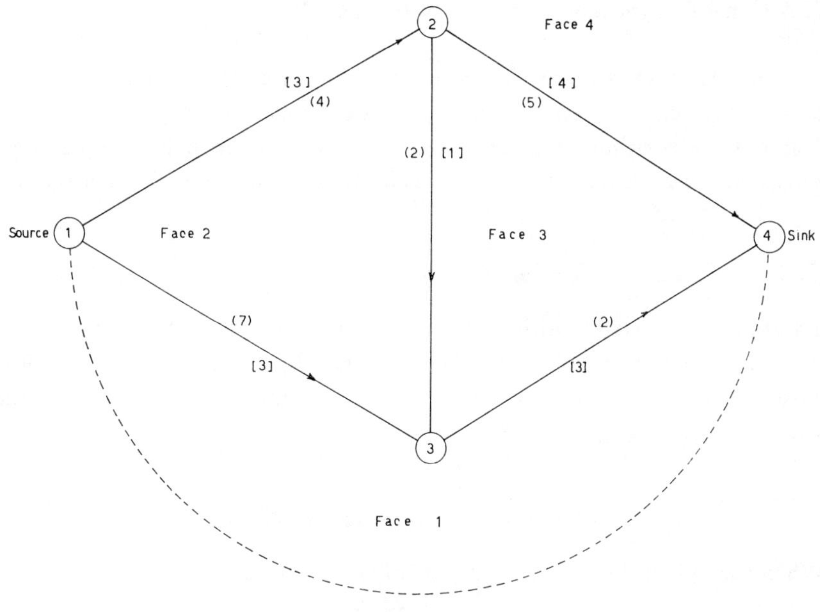

Fig. 6.2

branch costs. Table 6.1 gives the sequence of node and face numbers generated by the basic algorithm. Observe that the algorithm stops at step (a) of iteration 4 with a flow of 6 that corresponds to the maximum flow in this network.

TABLE 6.1

Iteration	Node Numbers				Face Numbers			
	1	2	3	4	1	2	3	4
0	0	0	0	0	0	0	0	0
1	0	4	6	8	0	0	1	1
2	0	4	7	9	0	2	3	5
3	0	5	7	10	0	3	3	6
4	0	∞	∞	∞				

7. A General Algorithm for Planar Networks

This section shows how the basic algorithm of section 6 can be used for solving the general piecewise linear convex problem on a planar network. The basic algorithm required initial vectors of node and face numbers consistent with all but one characteristic. In general there will be a set, say $N(u, V)$, of branches such that

$$(V_L - V_K,\ u_j - u_i) \notin C_{ij} \Leftrightarrow ij \in N(u, V). \tag{7.1}$$

To apply the basic algorithm one needs an extension of definition 3.1 for the length of the branches in N. Let $(x, y) = (V_L - V_K,\ u_j - u_i)$ be the flow tension pair of branch ij. When (x, y) is in the upper-left part of the plane set:

$$d_{ij}(u, v) = 0; \quad d_{ji}(u, v) \text{ as by def. 3.1}$$

$$d_{LK}(u, v) = 0; \quad d_{KL}(u, v) \text{ as by def. 3.1.}$$

When (x, y) is in the lower right part of the plane set:

$$d_{ji}(u, v) = 0; \quad d_{ij}(u, v) \text{ as by def. 3.1}$$

$$d_{KL}(u, v) = 0; \quad d_{LK}(u, v) \text{ as by def. 3.1.}$$

When (x, y) is on the characteristic use def. 3.1 (see fig. 7.1). The general algorithm as presented hereafter can be considered as a transposition to planar networks of the "out of kilter" algorithm of Ford and Fulkerson [6].

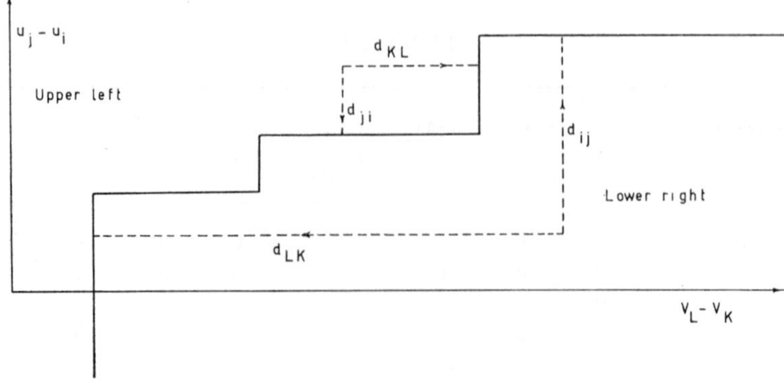

Fig. 7.1

THE GENERAL ALGORITHM

— Initialization: set $u_j^0 = 0$ $\quad j = 1, 2, \ldots, m$

$u_k^0 = 0$ $\quad k = 1, 2, \ldots, m$

$h = 1$

— Phase h

if $(u_j^{h-1} - u_i^{h-1}, V_L^{h-1} - V_K^{h-1}) \in C_{ij}$ all ij

END (*optimal solution*)

Otherwise let $(x_{ij}^{h-1}, y_{ij}^{h-1}) \notin C_{ij}$

(i) if (x_{ij}, y_{ij}) is in the upper-left part of the plane apply basic algorithm using (ij) as special branch

(a) optimal solution found go to phase $h+1$

(b) (*no feasible solution*) STOP

(c) (*unbounded objective*) STOP

(ii) if (x_{ij}, y_{ij}) is in the lower-right part of the plane apply basic algorithm using (ji) as special branch

(a) optimal solution found, go to phase $h+1$

(b) (*no feasible solution*) STOP

(c) (*unbounded objective*) STOP

References

1. BERGE, C. *The Theory of Graphs and their Application.* Methuen, London, 1962.
2. BERGE, C. and GHOUILA-HOURI, A. *Programming, Games and Transportation Networks.* Wiley, New York, 1965.
3. DE GHELLINCK, G. Aspects de la motion de dualité en théorie des graphes. *Cahiers du Centre d'Études de Recherche Opérationnelle*, **3** (No. 2), 1961.
4. DIAZ, H. and RAMINEZ, O. Un algorithme pour le problème du flot maximum dans un graphe planaire. Fac. Sc. Appl. Univ. Louvain, 1970.
5. DIJKSTRA, E. W. A note on two problems in connection with graphs. *Numerische Mathematik*, **1**, 1959.
6. FORD, L. R. and FULKERSON, D. R. *Flow in Networks.* Princeton University Press, 1962.
7. HU, T. C. *Integer Programming and Network Flows.* Addison-Wesley, Reading, 1969.

8. IRI, M. *Network Flows, Transportation and Scheduling, Theory and Algorithms.* Academic Press, New York, 1969.
9. MINTY, G. J. Monotone networks. *Proc. R. Soc.* (London) A, vol. 257, 1960.
10. MINTY, G. J. Solving steady-state nonlinear networks of monotone elements. *IRE Trans. CircuitTheory*, CT 8, 1961.
11. NEMHAUSER, G. A generalized permanent-label setting algorithm for the shortcut path between specified nodes. *CORE DP* No. 7005, Louvain, 1970.
12. WHITNEY, H. Planar graphs. *Fundamenta Math.*, **21**, 1933.

CHAPTER 9

TRANSPORTATION NETWORKS WITH RANDOM ARC CAPACITIES

P. DOULLIEZ and E. JAMOULLE

Summary

The arc capacities of a transportation network are assumed to be independent discrete random variables and the flow requirements at the sink nodes are known values. This paper presents an efficient method for finding the probability that all flow requirements be satisfied and the probability that a given arc be found in a minimal cut. The method is based on a decomposition principle which allows the transformation of the network capacity state space into nonoverlapping subsets. The expected amount of unsupplied flow is also found. The computational efficiency and domain of application of the method are discussed.

1. Introduction

The nodes of a transportation network are joined by arcs and the capacity of an arc is an upper bound to the flow that may pass over it. The flow must be sent through the network from "supply" nodes to "demand" nodes. The largest amount of flow that can be sent from a supply node is considered as the capacity of a fictitious arc joining the supply node and a common fictitious supply node, denoted as S. In this paper, the amounts of flow required at the different demand nodes i are known values d_i.

It is assumed that the arc capacities are independent discrete random variables. Hence, the probability that all flow requirements be satisfied is well-defined and will be computed by an efficient method. The probability that a given arc be found in a minimal cut will also be computed.[1] The

[1] Since the network has several sink nodes the definition of a minimal cut is here slightly different from the definition in [3]. We define a minimal cut as a set of arcs separating the common supply node S from one or several demand nodes i such that no supplementary flow can go from S to one of these demand nodes i.

method is based upon a decomposition principle which allows the transformation of the network capacity state space into nonoverlapping subsets.

In section 2 the decomposition principle is presented and is applied to our specific problem. It is shown in section 3 how the expected amount of unsupplied flow and the probability that a given arc be found in a minimal cut can be computed. Computational efficiency and domain of application are discussed in section 4.

2. The Decomposition Principle

2.1. *Notations and Definitions.* Let H_j be a discrete random variable ($j=1, \dots m$). The m random variables H_j are assumed to be independent. A random variable H_j assumes positive values $h_{j1}, h_{j2}, \dots h_{jk_j}$ with probabilities $p_{j1}, p_{j2}, \dots p_{jk_j}$ respectively. A point x of the state space can be defined as an m-tuple of values $x=(h_{1v_1}, h_{2v_2} \cdots h_{mv_m})$ where v_j is a numerical index for j going from 1 to k_j. For notational convenience, the index v_j will be used to designate the value h_{jv_j} so that a state point x will be denoted as $x=(v_1, v_2 \dots v_m)$. The entire state space is denoted by X. The state points $x=(1, 1, \dots 1)$ and $x=(k_1, k_2, \dots k_m)$ are called "limiting state points" for X.

In our specific problem, H_j is a capacity variable for arc j and h_{j1}, $h_{j2} \dots h_{jk_j}$ are capacity values for arc j. For simplicity we assume that the capacity values h_{jv_j} are such that $h_{j1}<h_{j2}< \dots <h_{jk_j} (j=1, \dots m)$. A state point $x=(v_1, v_2 \dots v_m)$ is called a network state.

Let $\Pr[X=x]$ be the probability associated with a point $x \in X$. We have:

$$\Pr[X=x] = p_{1v_1} \cdot p_{2v_2} \cdots p_{mv_m}$$

$$\sum_{v_j=1}^{k_j} p_{jv_j} = 1 \qquad (j=1, \dots m)$$

$$\sum_{\text{over } x} \Pr[X=x] = \Pr[X] = 1 .$$

2.2. *Statement of the Problem.* Suppose that any state point $x=(v_1, \dots v_m)$ of the state space X can be recognized as either acceptable or not. Let A^0 be the set of acceptable points and B^0 the set of non-acceptable points. The problem is to compute in an efficient way the probabilities $\Pr[A^0]$ and $\Pr[B^0]$. We have:

$$X = A^0 \cup B^0$$

$$\Pr[X] = \Pr[A^0] + \Pr[B^0] = 1 .$$

In our specific problem, a network state $x=(v_1, \dots v_m)$ is acceptable if the capacities v_j are such that all demands d_i can be satisfied. The probability $\Pr[A^0]$ is the probability that all demands be satisfied and $\Pr[B^0]$ is the probability that at least one demand be not entirely satisfied.

The state space X has $\prod_{j=1}^{m} k_j$ elements, so it is hopeless to solve the problem by considering each point of the state space one after the other. Dealing with subsets in which all state points are acceptable and subsets in which all state points are not acceptable, it can be seen in [1], p. 44 that the problem is still quite complex when these subsets may overlap. Therefore, we now present a method for constructing sequences of nonoverlapping subsets in which all state points are acceptable and nonoverlapping subsets in which all state points are not acceptable.

2.3. *Classification Rule.* Suppose that a state point $x=(v_1, \dots v_m)$ is acceptable whenever each v_j is within the range $[V_j^0, k_j]$ where V_j^0 is a known critical value for the random event H_j. Thus, $x=(v_1, \dots v_m)$ is an acceptable state point if $v_j^0 \le v_j \le k_j (j=1, \dots m)$.

Suppose also that a state point $x=(v_1 \dots v_j \dots v_m)$ is not acceptable whenever at least one v_j is within the range $[1, v_j^*[$ where v_j^* is a known critical value. Therefore, $x=(v_1 \dots v_j \dots v_m)$ is a non-acceptable state point if $1 \le v_j < v_j^*$ for at least one j.

If the classification rule holds, we can say that a state point $x=(v_1 \dots v_j \dots v_m)$ is "unspecified" whenever $v_j \ge v_j^*$ for all j and at least one v_j is within the range $[v_j^*, v_j^0[$, i.e. $v_j^* \le v_j < v_j^0$. It is explained in section 3 how the critical capacity values v_j^0 and v_j^* are obtained in our specific problem.

2.4. *Nonoverlapping Subsets of X.* In figure 1, a state point $x=(v_1, \dots v_m)$ can be represented by a line through the v_j table $(j=1, \dots m; m=5)$ and regions V_A, V_B, V_C are defined by the critical values v_j^* and v_j^0 as follows:

$$v_j \in V_A \quad \text{if} \quad v_j^0 \le v_j \le k_j$$

$$v_j \in V_B \quad \text{if} \quad 1 \le v_j < v_j^*$$

$$v_j \in V_C \quad \text{if} \quad v_j^* \le v_j < v_j^0 .$$

Classifying the state points $x=(v_1, v_2 \dots v_j \dots v_m)$ with respect to the critical values v_j^0 and v_j^*, it can be seen easily that any state point necessarily belongs to one of the following classes:

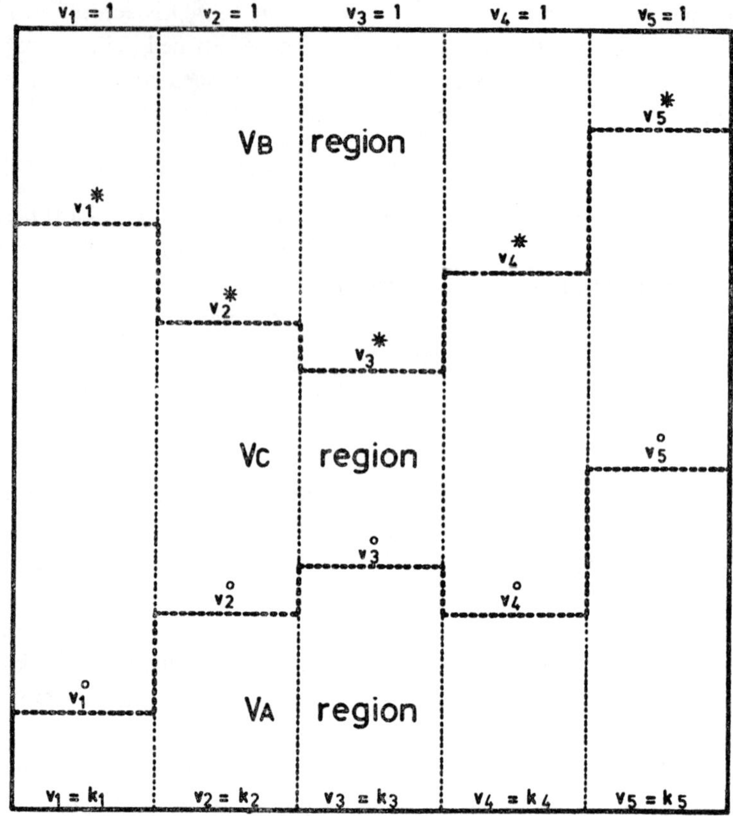

Fig. 1. The v_j index table ($m = 5$).

— x has its values v_j entirely in V_A. Then we say that $x \in A$ where A is a set of acceptable state points $x (A \subseteq A^0)$.
— x has at least one v_j in V_B. Then we say that $x \in B$ where B is a set of unacceptable state points $x (B \subseteq B^0)$.
— x has at least one v_j in V_C and no value in V_B. Then we say that $x \in C$ where C is a set of unspecified state points x.

We have:

$$A \cup B \cup C = X$$

$$A \cap B = \emptyset$$

$$B \cap C = \emptyset$$

$$A \cap C = \emptyset.$$

The sets B and C can be defined as follows:

$$B = \bigcup_{j=1}^{m} B_j$$

$$C = \bigcup_{j=1}^{m} C_j$$

where $x = (v_1, \dots v_m) \in B_l$ if $v_l < v_l^*$ and $x = (v_1, \dots v_m) \in C_l$ if $v_l < v_l^0$ and if $v_k \geq v_k^*$ $(k = 1, \dots m)$. The subsets B_j and C_j are represented in figure 2. Any two sets B_j (or C_j) may overlap. There is a one-to-one correspondance between a state x (represented as a line through the table in fig. 1) and a point of the set in figure 2 $(m = 5)$.

Since computing probabilities associated with overlapping subsets B_j and overlapping subsets C_j is meaningless, let us transform the subsets B_j and

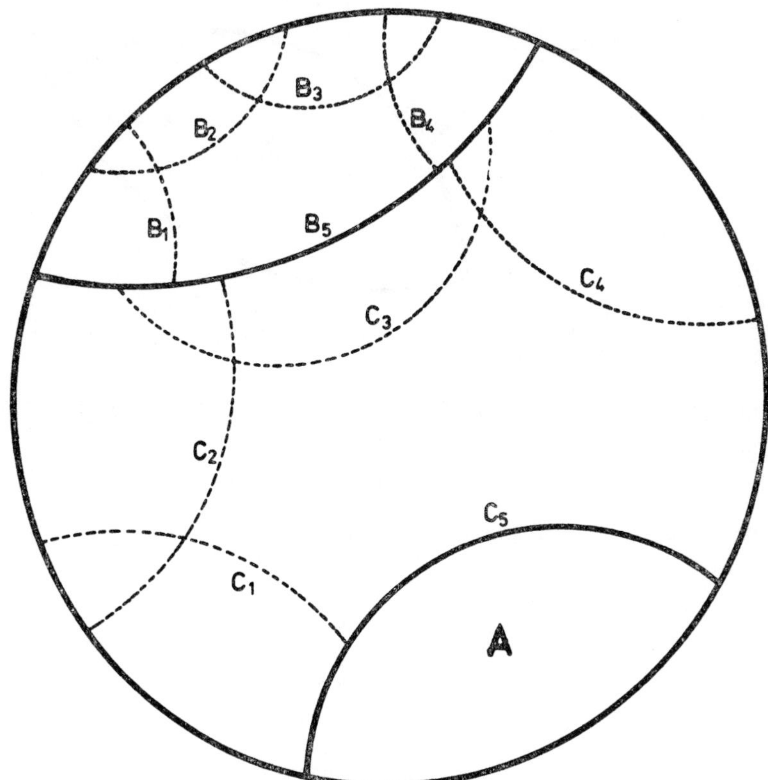

Fig. 2. The state space (overlapping subsets).

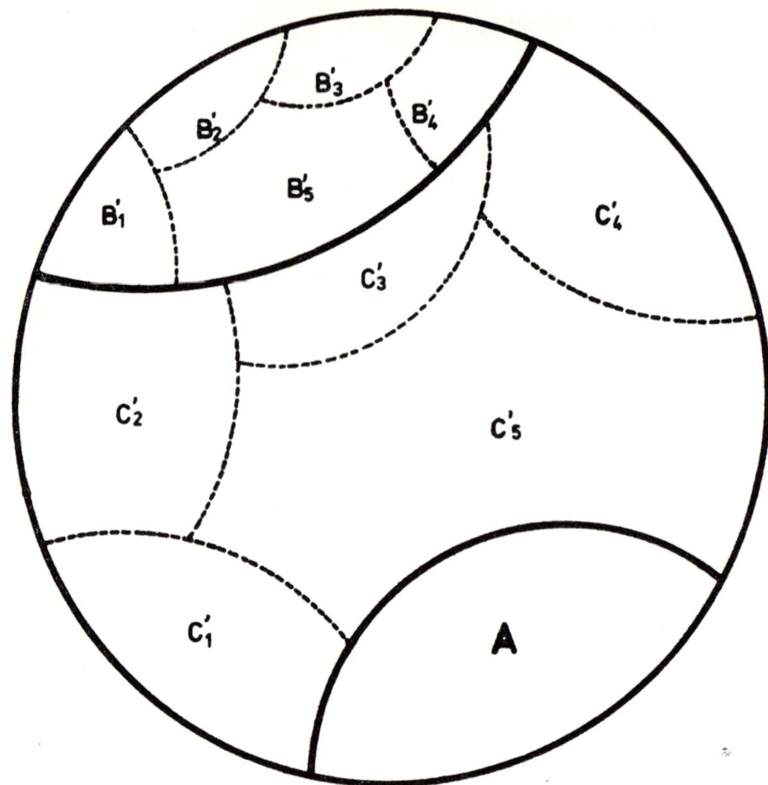

Fig. 3. The state space (nonoverlapping subsets).

C_j into B'_j and C'_j as represented in figure 3 such that for any j and $k (j \neq k)$ we have:

$$B'_j \cap B'_k = \emptyset \quad \text{and} \quad C'_j \cap C'_k = \emptyset .$$

The transformation of subsets $B_j(C_j)$ into subsets $B'_j(C'_j)$ can be performed sequentially as follows:

$$\begin{array}{ll}
B'_1 \equiv B_1 & C'_1 \equiv C_1 \\
B'_2 = B_2 - B'_1 \cap B_2 & C'_2 = C_2 - C'_1 \cap C_2 \\
\vdots & \vdots \\
B'_j = B_j - \sum\limits_{r<j} (B'_r \cap B_j) & C'_j = C_j - \sum\limits_{r<j} (C'_r \cap C_j) \\
\vdots & \vdots \\
B'_m = B_m - \sum\limits_{r<m} (B'_r \cap B_m) & C'_m = C_m - \sum\limits_{r<m} (C'_r \cap C_m) .
\end{array}$$

This can be easily translated in terms of v_r, v_l and cardinal value of index l for subsets B'_l and C'_l ($l = 1, \dots m$):

$x = (v_1, \dots v_m) \in B'_l$ if $v_l < v_l^*$ and $v_r \geq v_r^*$ for any $r < l$

$x = (v_1, \dots v_m) \in C'_l$ if $v_l^* \leq v_l < v_l^0$; $v_r \geq v_r^0$ for any $r < l$; $v_r \geq v_r^*$ for any $r > l$.

Thus, overlapping subsets B_j and C_j are transformed into nonoverlapping subsets B'_j and C'_j only by following the increasing order of indices j. It is easily seen that $B'_j \cap B'_k = \emptyset$ and $C'_j \cap C'_k = \emptyset$ for $j \neq k$. Also, $B'_j = \emptyset$ if $v_j^* = 1$ and $C'_j = \emptyset$ if $v_j^0 = v_j^*$.

2.5. *Probability Computations.* Since all sets A, B'_j and C'_j are exhaustive and nonoverlapping subsets of X, we have:

$$\text{(1)} \qquad \Pr[X] = \Pr[A] + \sum_{j=1}^{m} \Pr[B_j] + \sum_{j=1}^{m} \Pr[C_j].$$

It is easy to compute each term of the right-hand side in (1) since limiting state points are known for each subset, as indicated in figure 4.

If we define for each $j (j = 1, \dots m)$

$$p_j = \sum_{v_j = v_j^0}^{k_j} p_{jv_j}; \qquad q_j = \sum_{v_j = v_j^*}^{v_j^0 - 1} p_{jv_j}; \qquad s_j = \sum_{v_j = 1}^{v_j^* - 1} p_{jv_j}$$

we have

$$\text{(2)} \qquad \Pr[X] = \prod_{j=1}^{m} (p_j + q_j + s_j)$$

$$\text{(3)} \qquad \Pr[A] = \prod_{j=1}^{m} p_j$$

$$\text{(4)} \qquad \Pr[C'_j] = q_j \cdot \prod_{r=1}^{j-1} p_r \cdot \prod_{r=j+1}^{m} (p_r + q_r) \qquad j = 1, \dots m$$

$$\text{(5)} \qquad \Pr[B'_j] = s_j \cdot \prod_{r=1}^{j-1} (p_r + q_r) \cdot \prod_{r=j+1}^{m} (p_r + q_r + s_r) \qquad j = 1, \dots m.$$

The value $q_j = 0$ if $v_j^0 = v_j^*$ and $s_j = 0$ if $v_j^* = 1$.

In the last two equations, ignore the product $\prod_{r=a}^{b}$ whenever $b < a$. From equations (2) to (5) it can be shown algebraically that (1) holds.

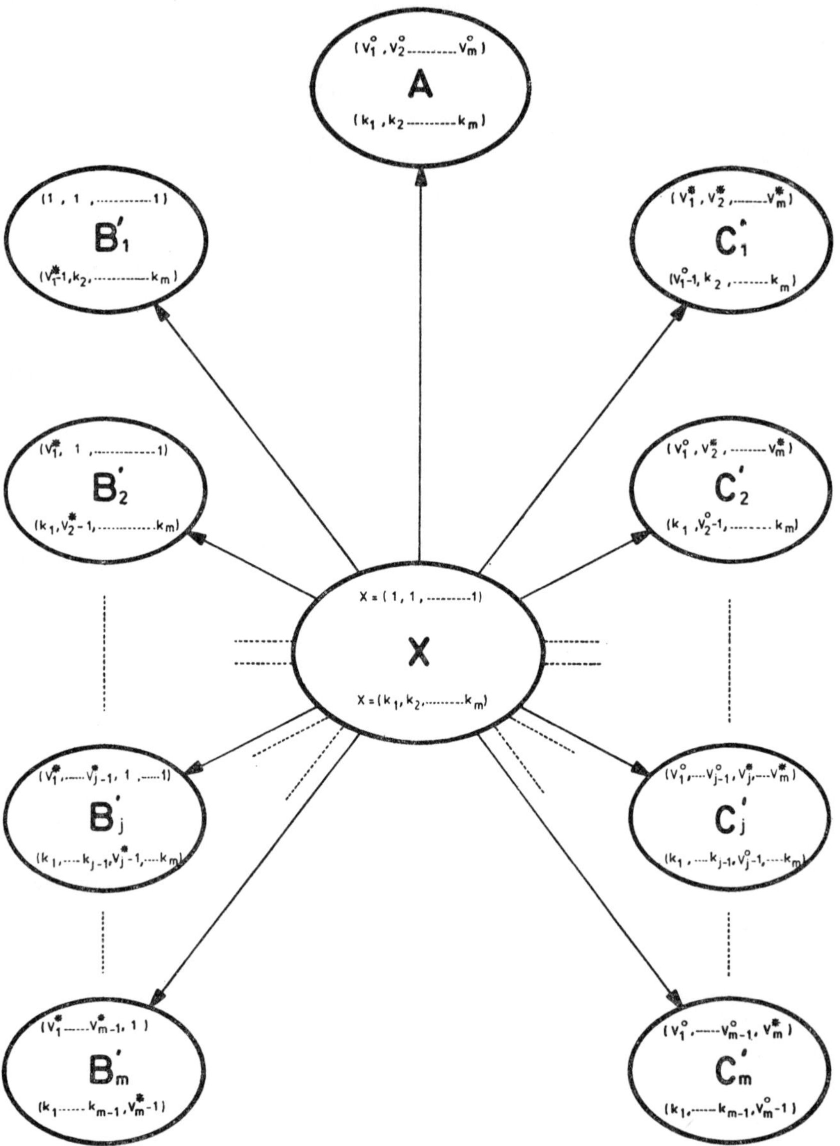

Fig. 4. The subsets of X and their limiting state points.

2.6. *The Decomposition Principle.* If values v_j^0 and v_j^* can be known for the initial state space X, nonoverlapping subsets A, B_j' and C_j' can be constructed. Any $x \in A$ is an acceptable state point and any $x \in B_j'$ $(j = 1, \dots m)$ an unacceptable state point. Any $x \in C_j'$ $(j = 1, \dots m)$ is an unspecified state point. Therefore, the associated probability Pr[A] is a part of the sought value $\Pr[A^0]$ and $\sum_{j=1}^{m} \Pr[B_j']$ is a part of the sought value $\Pr[B^0]$.

Now each nonempty subset C_j' may be considered as a new initial state space with modified but known limiting state points. In each C_j' new critical values v_j^0 and v_j^* may be determined, as they have been determined in the initial state space X. Thus, each C_j' may be again decomposed into new nonoverlapping subsets B_j' that contain only unacceptable points, a set A of acceptable points and new subsets of unspecified points. The probabilities associated with new sets A and B_j' can be computed as in (3) and (5) and are added to the former values and so on... until no subset with unspecified points can be generated and all such subsets have been considered. When this occurs, the probabilities $\Pr[A^0]$ and $\Pr[B^0]$ are obtained. The set A^0 and B^0 are, respectively, the union of all sets A and sets B_j' that have been constructed.

The procedure is finite. For an initial nonempty state space X, we suppose that A^0 is nonempty otherwise the problem is solved with $\Pr[A^0] = 0$. After decomposing X into A, B_j' and C_j' we have $A \subseteq X$ but if $A = X$, $\Pr[A^0] = 1$ and the problem is solved. Thus, $A \subset X$ and also each $B_j' \subset X$ and $C_j' \subset X$.

Similarly, at any stage of the procedure, each subset of unspecified state points is strictly contained into the set from which it has been generated. Consequently, the procedure is finite.

The procedure could be summarized as follows:

— Step 1: Start with initial set X. Let $\Pr[A^0] = \Pr[B^0] = 0$.

— Step 2: Given the limiting state points of X, compute the critical values v_j^0 and v_j^*. The sets A, B_j', C_j' can then be defined with their own limiting state points. Compute $\Pr[A]$ and $\Pr[B_j']$ for each j. Let $\Pr[A^0] = \Pr[A^0] + \Pr[A]$ and $\Pr[B^0] = \Pr[B^0] + \sum_{j=1}^{m} \Pr[B_j']$.

— Step 3: Choose a set C_j' that has not yet been considered. If no such set exists, go to step 4. Otherwise, let $X = C_j'$ and go to step 2.

— Step 4: $\Pr[A^0]$ and $\Pr[B^0]$ are the sought values. We necessarily have $\Pr[A^0] + \Pr[B^0] = 1$. END.

3. Transportation Network with Random Arc Capacities

3.1. *Critical Capacity Values.* In this section a method for finding critical values v_j^0 and v_j^* is presented for our specific network problem. They will be found with respect to initial network state set X with limiting state points $x=(1, 1, \ldots 1)$ and $x=(k_1, \ldots k_m)$ but the procedure would be the same for any set of network states with known limiting state points. The probability $\Pr[A]$ and probabilities $\Pr[B_j']$ can be computed once the critical capacity values v_j^0 and v_j^* have been determined for each random capacity H_j.

Let us join each demand node i to a common fictitious demand node T by way of an arc with capacity d_i. A flow $D=\sum_{\text{over } i} d_i$ is sent from the common source node S to T and the flow f_j in any arc j cannot be higher than k_j.

The values v_j^0 are obtained as follows: Let v_j^0 be the smallest capacity value that can be assumed by the discrete random variable H_j and which is not less than f_j. The network state $x=(v_1^0 \ldots v_j^0 \ldots v_m^0)$ is obviously acceptable. The set A is a set of network states $x=(v_1 \ldots v_j \ldots v_m)$ with $v_j^0 \leq v_j \leq k_j$ for any j. Therefore, any $x \in A$ is acceptable since it has been assumed in the previous section that capacity values are increasing with their indices.

The jth value v_j^* is obtained as follows: the set B_j' is a set of unacceptable network states $x=(v_1 \ldots v_j \ldots v_m)$ with $1 \leq v_j < v_j^*$ for arc j.

Given the limiting capacity values $k_r (r \neq j)$ and the existing flow D in the network, find the maximal flow F_j that can still go from the origin to the end of arc j without using arc j. The value F_j can be obtained by the Ford-Fulkerson method [3] and corresponds to the largest feasible decrease in flow for arc j. If $F_j < f_j$ arc j is in a minimal cut separating node S from node T and v_j^* is the smallest capacity value which is not less than $(f_j - F_j)$. Obviously, any $x \in B_j'$ is not able to satisfy all demands and is unacceptable. If $F_j \geq f_j$, then $v_j^* = 1$ and $B_j' = \emptyset$ since the entire range of capacity values for arc j is feasible for values d_i.

3.2. *Presence of an Arc in a Minimal Cut.* We are now interested in finding the probability $\Pr[B^0 | r]$ that arc r be found in a minimal cut $(r=1, \ldots m)$. Since any unacceptable network state has a minimal cut associated with it, only those with a minimal cut that contains arc r must be retained and their associated probabilities must be summed up.

For any set of unacceptable network states B_j' that has been generated by the procedure explained in section 2, let us make the following decomposition. The limiting state points of a set B_j' are known and let us designate them as $x=(g_1, \ldots g_m)$ and $x=(k_1, \ldots k_m)$ with $g_j \leq k_j$ for all j. Let M be a minimal cut associated with a network state x that is defined by the highest capacity values, i.e. $x=(k_1, \dots k_m)$. The value of the cut M is necessarily lower than the flow D which must go from S to T but cut M is not necessarily a minimal cut for network states with capacity values $g_j \leq v_j \leq k_j$. The decomposition principle must now be applied to the set B'_j with a different criterion, i.e. B'_j must be divided into (nonoverlapping) subsets of network states with identical minimal cut.

Let $X=B'_j$. The set X can be decomposed into a subset A and new subsets $B'_j(j=1, \dots m)$ when critical capacity values $(v_1^0 \dots v_j^0 \dots v_m^0)$ are determined; we have:

$$x=(v_1 \dots v_m) \in A \quad \text{if} \quad v_j \geq v_j^0 \quad \text{for all } j\,,$$

$$x=(v_1 \dots v_m) \in B'_j \quad \text{if} \quad v_j < v_j^0 \quad \text{for at least one } j\,.$$

The critical values v_j^0 are obtained as follows. For any arc $j \notin M$, let v_j^0 be the smallest capacity value which is not less than the flow value in arc j. For any arc $j \in M$, let $v_j^0 = g_j$.

Obviously, any network state $x \in A$ has the minimal cut M associated with it. If arc r is in cut M, the value $\Pr[A]$ is a part of the sought value $\Pr[B^0|r]$ and is to be added to the part already computed. The network states x in a new set B'_j do not necessarily have an identical minimal cut. Therefore, each new set B'_j is to be decomposed in the same manner. The value $\Pr[B^0|r]$ is obtained when no subset remains for decomposition.

3.3. *Existence of a Minimal Cut.* It is interesting to note that probability $\Pr[B^0|L]$ that a cut L be minimal can be found similarly by summing up all probabilities associated with sets in which the network states have L as minimal cut. Thus the problem that has been formulated in [4] by Frank and Hakimi can be solved by our approach.

3.4. *Expected Value of Unsupplied Flow.* Let A be a set of unacceptable network states with identical minimal cut and with known limiting state points. It has been shown above how an exhaustive sequence of sets A can be generated. Let the limiting state points of A be $x=(g_1 \dots g_j \dots g_m)$ and $x=(k_1 \dots k_j \dots k_m)$ with $k_j \geq g_j$ for all j. Let D be the total demand and $G(x)$ the value of the minimal cut for state $x \in A$.

Each network state $x \in A$ gives rise to an unsupplied flow $(D-G(x))$ with a probability $\Pr[A=x]=p_x$. The expected value $\mu[A]$ of the unsupplied flow associated with A is

$$\mu[A] = \sum_{x \in A} p_x(D-G(x)) = \Pr[A] \cdot D - \sum_{x \in A} p_x G(x)\,.$$

The term $\sum_{x \in A} p_x G(x)$ is the expected value of minimal cut M within set A.

Let

$$q_j = \sum_{v_j = g_j}^{k_j} p_{jv_j}$$

$$\bar{h}_j = \sum_{v_j = g_j}^{k_j} h_{jv_j}$$

and be M the minimal cut for any $x \in A$.

From the fact that the expected value of a sum of independent random variables is the sum of the expected values of these variables, we have:

$$\sum_{x \in A} p_x G(x) = \prod_{j \notin M} q_j \Big(\sum_{j \in M} \bar{h}_j \cdot \prod_{\substack{r \in M \\ r \neq j}} q_r \Big)$$

$$= \sum_{j \in M} \bar{h}_j / q_j \cdot \Pr [A] .$$

Therefore

$$\mu[A] = \Pr [A] \cdot (D - \sum_{j \in M} \bar{h}_j / q_j) . \tag{1}$$

Computing the right-hand side of (1) is straightforward. Summing up the values in (1) for all generated sets A gives the expected value of the unsupplied flow. Similarly, summing up the values in (1) for generated sets A for which the minimal cut contains arc r gives the expected value $\mu[A \mid r]$ of the flow which is unsupplied when arc r is present in a minimal cut. Finally, summing up the values in (1) for generated sets A for which cut M is minimal for any $x \in A$ gives the expected value $\mu[A \mid M]$ of the flow which is unsupplied when cut M is minimal.

4. Concluding Remarks

The decomposition principle presented in this paper is general in the sense that it could be applied in any other system in which a point of the state space is defined when several random events occur simultaneously and when a point of the state space can be classified according to a given criterion.

The problem of finding the probability of meeting the demands at the nodes of a network has been treated thus far by performing a large number of successive random trials (Monte-Carlo approach). The method presented in this paper finds an exact answer to the problem with a small computational effort. Anyway, the Monte-Carlo approach is not appropriate for estimating properties of a set of state points which is reached very rarely even after a great number of random trials. For some networks that have been treated by our approach, the probability associated with all unacceptable state points was not higher then 0.001. A main reason for efficiency is that a sequence of nonoverlapping sets of network states are considered instead of a sequence of single network states as in the Monte-Carlo approach.

When the demands at the demand nodes are increasing with time, the method can be easily adapted for finding the largest value of time up to which all demands can be satisfied with a given probability level. Also, a best location for an investment on a network can be found since the responsibility of each arc for not satisfying the demands is evaluated with respect to the entire network.

Several numerical examples have been treated by a computer program written in FORTRAN IV. With the present limitations of the program, the number of arcs cannot exceed 100 and the number of nodes cannot exceed 50.

The present paper can be considered as a probabilistic extension of reference [2]. In [2] the required probability level with which all demands must be satisfied has been taken into account only in an empirical way, i.e. all demands must be satisfied even if any one arc assumes a given lower capacity.

References

1. DOULLIEZ, P. Probability distribution function for the capacity of a multiterminal network. *R.I.R.O.*, **1**: 39–49, 1971.
2. DOULLIEZ, P. and RAO, M. R. Maximal flow in a multiterminal network with any one arc subject to failure. *Management Science*, **18**, 1971.
3. FORD, L. R. and FULKERSON, D. R. *Flows in Networks*. Princeton University Press, Princeton, New Jersey, 1962.
4. FRANK, H. and HAKIMI, S. L. Probabilistic flow through a communication network. *IEEE Trans. on Circuit Theory*, 413–414, 1965.

CHAPTER 10

AN OPTIMALITY CONDITION FOR MINIMAL-COST MULTICOMMODITY FLOWS

IRINEL DRAGAN

1. Introduction

Consider a network R defined by: a graph $G=(N, U)$, $|N|=m$, $|U|=n$; p supply functions $b^h(x_i)$, $\forall x_i \in N$, $(h=1, \ldots, p)$; a capacity function $d(u_k) \geq 0$, $\forall u_k \in U$, $(k=1, \ldots, n)$. The graph G, which is a directed finite graph without loops, is given by its incidence matrix A. The supply functions are given as m-vectors and we shall denote by B the matrix which has these vectors as columns. We suppose that the supplies are integers and the sum of all supplies in the network equals zero for each $h=1, \ldots, p$. The capacity function is given as an n-vector D and we suppose that the capacities are integer numbers, too. A multicommodity flow in R is an $n \times p$ matrix $X=(X^1, \ldots, X^p)$ satisfying

$$\begin{cases} AX = B, & (1) \\ X \cdot 1 \leq D, & (2) \\ X \geq 0, & (3) \\ X = \text{integer}, & (4) \end{cases}$$

where 1 is a p-vector $1=(1, \ldots, 1)'$. The last condition has been introduced in our definition for simplicity.

Consider the case where the cost function $c_k(x_k^1, \ldots, x_k^p)$, $\forall u_k \in U$, $(k=1, \ldots, n)$, is linear,

$$c_k(x_k^1, \ldots, x_k^p) = \sum_{h=1}^{p} c_k^h x_k^h, \qquad (k=1, \ldots, n), \tag{5}$$

which means that it can be given by an $n \times p$ matrix $C=(C^1, \ldots, C^p)$. Then the total cost of the multicommodity flow X in R is

$$C(X) = \sum_{h=1}^{p} (C^h)' \cdot X^h. \tag{6}$$

The problem that concerns us below is: (P) minimize the function (6) under the constraints (1)–(4). Comprehensive references are given in [3] for the case $p=1$, but the general problem has been often considered only in the last years.

In the single-commodity case a well-known theorem represents a necessary and sufficient condition for optimality: "a flow with value v is optimal if and only if, based on the modified costs, there exist no negative cycles with respect to the flow", (see [3], p. 170). This is a primal necessary and sufficient condition for optimality, because it does not use the dual elements of the problem. A constructive proof of the theorem has been given in [4], where an algorithm for solving the problem has been justified by means of this theorem. In a recent paper [2], we gave a sufficient condition for optimality in the multicommodity case with equal costs, by using an extension of a result from [4]. In the following a necessary and sufficient condition for optimality will be given in the general case. However, an algorithm based upon this result, and representing an extension of Menon's algorithm, seems to be inefficient.

2. Optimality Condition

Consider a multicommodity flow X in the network R. If V^* is a vector-cycle corresponding to a directed cycle in G, then it can be shown that $AV^*=0$. So, if $V=(V^1, \ldots, V^r)$ is an $n\times r$ matrix of vector-cycles, corresponding to a system of directed cycles in G, then we shall have $AV=0$. Denote by $\varepsilon=(\varepsilon^1, \ldots, \varepsilon^r)$ a $p\times r$ integer matrix and consider the matrix X^* determined by

(7) $$X^* = X + V\cdot\varepsilon'.$$

According to the remark above this matrix satisfies (1) and (4); if X^* satisfies (2) and (3), then it would be a multicommodity flow.

A system of directed cycles is called *conformal* if each arc belonging to several cycles is either a forward arc in all the cycles or a backward arc in all the cycles.

Now, we can state the basic result given in [2], as follows.

THEOREM 1. If X and $\overline{X}$ are multicommodity flows, then there exists a conformal system of cycles $V=(V^1, \ldots, V^r)$ and an integer matrix $\varepsilon=(\varepsilon^1, \ldots, \varepsilon^r)$ with $\varepsilon'\cdot 1\geq 0$, such that

(8) $$\overline{X} = X + V\cdot\varepsilon'.$$

This result gives a decomposition in terms of cycles for the difference of two multicommodity flows, because the compact formulas (8) can be written as

$$
\begin{aligned}
\bar{X}^1 &= X^1+\varepsilon_1^1 V^1+ \dots +\varepsilon_1^r V^r, \\
&\dots\dots\dots\dots\dots\dots \\
\bar{X}^p &= X^p+\varepsilon_p^1 V^1+ \dots +\varepsilon_p^r V^r.
\end{aligned} \tag{9}
$$

The inequality $\varepsilon' \cdot 1 \geq 0$, or

$$\sum_{h=1}^{p} \varepsilon_h^t \geq 0, \qquad (t = 1, \dots, r), \tag{10}$$

shows that the formulas (9) represent really an extension of a well-known result (see [1], p. 145 and [4], p. 165).

Now the relationship between the costs of the multicommodity flows X and $\bar{X}$ is

$$C(\bar{X}) = C(X) + \sum_{t=1}^{r} \sum_{h=1}^{p} \varepsilon_h^t \cdot [(C^h)' \cdot V^t]; \tag{11}$$

this equality will provide us with a necessary and sufficient condition for optimality.

A conformal system of cycles, given by the matrix $V=(V^1, \dots, V^r)$, will be called a feasible system of cycles with respect to the multicommodity flow X, if there exists a $p \times r$ integer matrix $\varepsilon = (\varepsilon^1, \dots, \varepsilon^r)$ with $\varepsilon' \cdot 1 \geq 0$ such that $X + V \cdot \varepsilon'$ be a multicommodity flow. Of course, if V gives a feasible system of cycles, then $X + V \cdot \varepsilon'$ could be a multicommodity flow for several $p \times r$ integer matrices ε with $\varepsilon' \cdot 1 \geq 0$; all these matrices will be called feasible with respect to the feasible system of cycles V. Now we can easily prove the following necessary and sufficient condition for optimality:

THEOREM 2. A multicommodity flow $\hat{X}$ is a minimum-cost multicommodity flow in R, if and only if for each feasible system of cycles $W = (W^1, \dots, W^r)$ and a feasible matrix $\eta = (\eta^1, \dots, \eta^r)$, we have

$$\sum_{t=1}^{r} \sum_{h=1}^{p} \eta_h^t \cdot [(C^h)' \cdot W^t] \geq 0. \tag{12}$$

The necessity is obvious and the sufficiency follows from theorem 1.

Let us give several comments:

(a) If $C^1 = C^2 = \ldots = C^p = \gamma$, as was supposed in [2], and we denote by

$$\lambda^t = \sum_{h=1}^{p} \eta_h^t, \qquad (t = 1, \ldots, r), \tag{13}$$

then we can rewrite (12) as

$$\sum_{t=1}^{r} \lambda^t (\gamma' W^t) \geq 0; \tag{14}$$

so, if

$$\gamma' \cdot W^t \geq 0, \qquad (t = 1, \ldots, r), \tag{15}$$

then (14) will be satisfied, because $\lambda^t \geq 0$, $(t = 1, \ldots, r)$. The sufficient condition for optimality proved in [2] required just the fulfillment of (15).

(b) In the single-commodity case, $(p = 1)$, our condition (12) becomes

$$\sum_{t=1}^{r} \eta^t (\gamma' W^t) \geq 0. \tag{16}$$

If W represents a feasible system of cycles, then it can be shown that W^t is a feasible cycle for each $t = 1, \ldots, r$; therefore, in the case $p = 1$ our theorem 2 is equivalent to the well-known theorem cited above from [3]. Unfortunately, in the multicommodity case it is possible to have a feasible system of cycles W, although W^t are not feasible cycles for each $t = 1, \ldots, r$; therefore, it is impossible to state our theorem 2 by using only the costs of these cycles, i.e. the numbers $(C^h)' \cdot W^t$, $(h = 1, \ldots, p;\ t = 1, \ldots, r)$.

(c) An algorithm for solving the minimum-cost multicommodity flow problem, based upon theorem 2, seems to be inefficient. Indeed, suppose that a subroutine for providing successively the system of cycles would be available. Then, in each step of the algorithm it would be necessary to establish for the corresponding matrix W whether the system

$$\begin{cases} (X + W \cdot \eta') \cdot 1 \leq D, \\ X + W \cdot \eta' \geq 0, \\ \sum_{t=1}^{r} \sum_{h=1}^{p} \eta_h^t \cdot [(C^h)' \cdot W^t] < 0, \end{cases} \tag{17}$$

is consistent or not. In the first case a better multicommodity flow would be $X^* = X + W \cdot \eta'$, and a new group of steps would improve eventually the new multicommodity flow X^*; in the second case a new step would continue the same search for another system of cycles. Moreover, it is difficult to generate successively all the systems of cycles.

References

1. Berge, C. and Ghouila-Houri, A. *Programmes, Jeux et Réseaux de Transport.* Dunod, Paris, 1962.
2. Dragan, I. An optimality criterion for minimal-cost multicommodity flows with proportional costs. *CORE Discussion Paper 7043*, Louvain, December 1970.
3. Hu, T. C. *Integer Programming and Network Flows.* Addison-Wesley, Reading, Massachusetts, 1969.
4. Menon, V. V. The minimal-cost flow problem with convex costs. *N.R.L.Q.*, **12**: 163–172, 1965.

CHAPTER 11

AN ALGORITHM FOR THE LOCATION OF CENTRAL FACILITIES

PIERRE HANSEN and LEON KAUFMAN

Summary

The constrained facilities-location problem (CFLP) studied in this paper may be stated as follows: consider an area which contains a given number of population centers and wherein an unknown number of public facilities, which are to serve these populations, must be located. Assume that a set of feasible locations for the facilities is given and that any given member of the public will always go to the nearest facility. Assume further that the establishment of facilities is limited by an investment constraint. Determine, with these assumptions, the number, the sizes and the locations of the facilities to be established in order to minimize the total transportation costs of the users.

An implicit-enumeration algorithm is proposed for solving the CFLP. First, the CFLP is shown to be equivalent to a sequence of simple plant location problems with two side conditions. Then, tight bounds on the values taken by the objective function and by the left-hand side of the investment constraint are obtained by using a theorem of Efroymson and Ray and the recent idea of additive penalties. These bounds provide efficient direct and conditional tests used in the algorithm. Computational experience on a CDC 6400 is reported and discussed.

1. Introduction

C. Revelle and P. Rogeski [10] have stated the following location-transportation problem: consider an area which contains a given number of population centers (i.e. towns, villages etc.) and wherein an unknown number of public facilities (i.e. hospitals, schools etc.), which are to serve these populations, must be located. Assume that a set of feasible locations for the facilities is given and that the public will always go to the nearest facility. Assume further that the establishment of facilities is limited by an investment constraint. Determine, under these assumptions, the number, the sizes and the locations of the facilities to be established in order to minimize total transportation costs for the users, i.e. the weighted sum of transportation distances and/or times.

Let us call this problem the *constrained facilities-location problem* (CFLP). The CFLP is related to other well-known location-transportation problems such as the *simple plant-location problem* [1], [2], [12], [13], [15] and the *p-median location problem* [4], [9], [11], [14]. This relation will be used below.

Revelle and Rogeski [10] have presented an algorithm for the CFLP based on linear programming. This algorithm provides optimal solutions for some (a priori unknown) values of the upper bound on the investment funds. Given such an upper bound, a feasible solution may be obtained by lowering the value of this bound according to the rules of the algorithm. The feasible solution thus obtained will be optimal for the new value of the upper bound, but it will not necessarily be optimal also for the original value. To the authors' knowledge, no general algorithm for the CFLP has yet been published. It is the object of this paper to derive such an algorithm.

In the next section a mathematical formulation of the CFLP is given. In section 3 the relation of this problem to the p-median location problem is explained. It is shown how tight bounds on the values taken by the objective function or the left-hand side of the investment constraint may be obtained by using the recent idea of *additive penalties* [5], [6]. These bounds are used in the implicit-enumeration algorithm given in section 4. Computational experience is discussed in section 5.

2. Mathematical Formulation

The CFLP may be stated as follows:

(1) $$\text{minimize} \sum_i \sum_j c_{ij} x_{ij}$$

under the constraints

(2) $$\sum_i g_i y_i + \sum_i \sum_j a_{ij} x_{ij} \leq b$$

(3) $$y_k c_{ij} x_{ij} \leq c_{kj}$$

(4) $\sum_i x_{ij} = 1$

(5) $x_{ij} \leq y_i$

(6) $x_{ij} \in \{0, 1\}$

(7) $y_i \in \{0, 1\} \qquad (i = 1, 2, ..., m;\ j = 1, 2, ..., n;\ k = 1, 2, ..., m)$,

where

c_{ij} is the transportation cost incurred by the public from the jth population center when going to the ith location,

g_i is a fixed investment cost for establishing a facility in the ith location,

a_{ij} is a proportional investment cost incurred if the public from the jth population center uses a facility established in the ith location,

b is an upper bound on investment funds,

x_{ij} is a boolean variable equal to 1 if the public from the jth population center uses a facility at the ith location, and equal to 0 otherwise,

y_i is a boolean variable equal to 1 if a facility is established in the ith location, and equal to 0 otherwise.

As stated above, the objective function (1) to be minimized is the sum of all transportation costs. The investment constraint (2) requires that the sum of the fixed and proportional investment funds expended does not exceed the upper bound b. Constraints (3) express the assumption that the public from each population center always goes to the nearest facility[1]. Indeed, consider the constraints (3) for a given value of k; if $y_k = 0$ there is no facility in the kth location and the constraints disappear; if $y_k = 1$ there is a facility in the kth location and the constraints reduce to

(8) $c_{ij} x_{ij} \leq c_{kj} \qquad (i = 1, 2, ..., m;\ j = 1, 2, ..., n)$,

which imply that no one can go to a facility located further from his population center than the kth location. The very numerous quadratic constraints (3) need not be written explicitly, but will be implicitly taken into account during the solution. The constraints (4) express that the public from each population center goes to a facility and constraints (5) that nobody goes for service to a location where there is no facility. Finally, constraints (6) express that the public of each population center either goes as a whole or

[1] Assuming transportation cost proportional to distance. Otherwise, it goes to the cheapest-to-reach facility.

does not go at all to any particular location; and constraints (7) express that no fractional facilities may be established.

Formulation (1)–(7) of the CFLP differs from that of Revelle and Rogeski in several respects: constraints (3) are explicit, m is not restricted to be equal to n and the objective function and investment constraint are more general than those proposed by those authors, because the coefficients c_{ij} and a_{ij} are not assumed to be directly proportional to the numbers of inhabitants of the population centers.

3. Bounds

Let us notice that, once the values of the variables y_i are known, the values of the variables x_{ij} are determined by constraints (3)–(6). This allows the resolution of the CFLP by implicit enumeration of the vectors of values of the y_i's. It is well-known that a crucial factor for an implicit enumeration algorithm to be computationally efficient is the possibility of obtaining tight bounds to be used in the tests. For the CFLP bounds must be obtained on the values taken by the objective function and on the values taken by the left-hand side of the investment constraint. These bounds will be obtained by using a well-known result of M.A. Efroymson and T.L. Ray [2] for the simple plant-location problem, elaborated in [6], [7], [8].

Notice that, if constraints (2) and (3) are omitted (or only constraint (3) is redundant when (2) is omitted) the CFLP reduces to the simple plant-location problem, with the difference, however, that no fixed costs appear in the objective function. A lower bound on the values taken by the objective function could be derived as in [2], but would probably be loose, because of the absence of fixed costs. However, if the number of facilities to be established is imposed, i.e. if a constraint of the form

$$\sum_i y_i = p \tag{9}$$

(where p is an integer between 1 and m) is added to the CFLP, the objective function may be replaced by the new objective function,

$$\text{minimize } f \sum_i y_i + \sum_i \sum_j c_{ij} x_{ij}, \tag{10}$$

where f is a positive constant (the value of which will be discussed in section 5). Indeed, the effect of this substitution is to increase the value of every feasible solution by the constant value $p.f.$ If the value of f is judiciously chosen, tight bounds on the values of the objective function (10) will be obtained. Using this substitution, the CFLP is replaced by a sequence of m problems in which the objective function is (10), the constraint (9) is added, and p takes the values $1, 2, \ldots, m$.

The problems of this sequence may be viewed as simple plant-location problems with two *side conditions* (i.e. (2) and (9)) or as p-median location problems with a side condition (i.e. (2)). Let us consider the pth problem of this sequence, at a current iteration of its resolution by an implicit enumeration algorithm. Let K_0 denote the set of indices of the y_i's fixed at 0, K_1 the set of indices of the y_i's fixed at 1, and K_2 the set of indices of the y_i's, the values of which have not yet been fixed:

$$K_0 = \{i \mid y_i = 0\} \tag{11}$$

$$K_1 = \{i \mid y_i = 1\} \tag{12}$$

$$K_2 = \{i \mid y_i \in \{0, 1\}, \text{ not yet fixed}\} \tag{13}$$

and let

$$m_0 = |K_0| \tag{14}$$

$$m_1 = |K_1| \tag{15}$$

$$m_2 = |K_2|. \tag{16}$$

The sets of indices K_0, K_1 and K_2 define a *partial solution* (see A. Geoffrion [3]). The assignment of boolean values to the variables whose indices are in K_2 (or *free variables*) defines a *completion* of this partial solution. The set of all completions of the current partial solution will be noted $I(K_0, K_1, K_2)$:

$$I(K_0, K_1, K_2) = \{Y \mid y_i = 0, \forall i \in K_0, y_i = 1, \forall i \in K_1, y_i \in \{0, 1\}, \forall i \in K_2\}. \tag{17}$$

Further, let $I'(K_0, K_1, K_2)$ denote the set of completions of $I(K_0, K_1, K_2)$ to which corresponds a feasible solution. The shortest distance between the jth population center and an already established facility will be noted c^*_{rj}:

$$c^*_{rj} = \min_{i \in K_1} c_{ij}. \tag{18}$$

Let n_i denote, for any $i \in K_2$, the number of population centers which are nearer to the ith location than to any already established facility:

$$n_i = |\{j \ni c_{ij} \leq c^*_{rj}\}|. \tag{19}$$

Further, let

$$(20)\qquad c'_{ij} = \begin{cases} c_{ij}+f/n_i & \forall i \in K_2 \\ c_{ij} & \forall i \in K_1 \end{cases}$$

and let c'_{rj} and c'_{sj} denote the first and second minima of the c'_{ij}'s for all $i \in K_1 \cup K_2$:

$$(21)\qquad c'_{rj} = \min_{i \in K_1 \cup K_2} c'_{ij}$$

$$(22)\qquad c'_{sj} = \min_{i \in K_1 \cup K_2,\, i \neq r} c'_{ij}.$$

At the current iteration considered, the pth problem may be written:

$$(23)\qquad \text{minimize } m_1 f + f \sum_{i \in K_2} y_i + \sum_{i \in K_1 \cup K_2} \sum_j c_{ij}$$

under the constraints

$$(24)\qquad \sum_{i \in K_1} g_i + \sum_{i \in K_2} g_i y_i + \sum_{i \in K_1 \cup K_2} \sum_j a_{ij} x_{ij} \leq b$$

$$(25)\qquad y_k c_{ij} x_{ij} \leq c_{kj} \quad \text{for } k \in K_2$$

$$(26)\qquad c_{ij} x_{ij} \leq c_{kj} \quad \text{for } k \in K_1$$

$$(27)\qquad m_1 + \sum_{i \in K_2} y_i = p$$

$$(28)\qquad \sum_{i \in K_1 \cup K_2} x_{ij} = 1$$

$$(29)\qquad x_{ij} \leq y_i$$

$$(30)\qquad x_{ij} \in \{0, 1\}$$

$$(31)\qquad y_i \in \{0, 1\} \qquad (i \in K_1 \cup K_2;\ j = 1, 2, \ldots, m;\ k \in K_1 \cup K_2).$$

After deletion of constraints (24) to (27) this problem has the same form as the simple plant-location problem. Efroymson and Ray's theorem shows that

$$(32)\qquad \underline{f} = m_1 f + \sum_j c'_{rj}$$

is a lower bound of the values taken by (23) on $I'(K_0, K_1, K_2)$.

A *penalty* p_i^0 (or p_i^1) is a quantity which may be added to $\underline{f}$ when a free variable is fixed at 0 (or at 1), thus yielding a valid new lower bound on the values taken by (23) on $I'(K_0 \cup \{i\}, K_1, K_2 \backslash \{i\})$

$$(\text{or } I'(K_0, K_1 \cup \{i\}, K_2 \backslash \{i\})).$$

If several variables are simultaneously fixed at 0 or at 1, and the addition of the corresponding penalties yields a valid new bound, these penalties are said to be additive; i.e. if $\forall K_0', K_1' \subset K_2 \ni K_0' \cap K_1' = \emptyset$:

$$f + \sum_{i \in K_1'} p_i^1 + \sum_{i \in K_0'} p_i^0 \leq f', \tag{33}$$

where f' is a lower bound of (23) on $I'(K_0 \cup K_0', K_1 \cup K_1', K_2 \backslash (K_0' \cup K_1'))$, the penalties p_i^0 and p_i^1 are *additive*.

As shown in (7), the penalties

$$p_i^0 = \sum_j \max(c_{sj}' - c_{ij}', 0) \tag{34}$$

and

$$p_i^1 = f - \sum_{j \mid r \in K_1 \cup \{i\}} \max(c_{rj}' - c_{ij}, 0) - \sum_{j \mid r \in K_2 \backslash \{i\}} \max(c_{sj}' - c_{ij}, 0) \tag{35}$$

are additive. As every free variable y_i must take the value 0 or 1 in any completion of the current partial solution,

$$f^1 = f + \sum_{i \in K_2} \min(p_i^0, p_i^1) \tag{36}$$

is a lower bound of (23) on $I'(K_0, K_1, K_2)$.

Further, let for every $i \in K_2$

$$d_i = |p_i^0 - p_i^1| \tag{37}$$

and let

$$K_3 = \{i \in K_2 \mid p_i^0 \geq p_i^1\} \tag{38}$$

$$K_4 = \{i \in K_2 \mid p_i^1 > p_i^0\} \tag{39}$$

$$m_3 = |K_3| \tag{40}$$

$$m_4 = |K_4|; \tag{41}$$

let d_k^0, $k = 1, 2, \ldots, m_3$ denote the d_i's the indices of which are in K_3, ranked in order of increasing value, and let d_k^1, $k = 1, 2, \ldots, m_4$ denote the d_i's the indices of which are in K_4, also ranked in order of increasing value.

As shown in (8), because of the constraint (27),

$$f^2 = \begin{cases} f^1 + \sum_{k=1}^{p-m_1-m_3} d_k^1 & \text{if} \quad m_1+m_3<p \\ f^1 & \text{if} \quad m_1+m_3=p \\ f^1 + \sum_{k=1}^{m_1+m_3-p} d_k^0 & \text{if} \quad m_1+m_3>p \end{cases} \tag{42}$$

is a lower bound of (23) on $I'(K_0, K_1, K_2)$.

Lower bounds on the values taken by the left-hand side of the investment constraint (24) may be obtained in a similar manner. Indeed the problem,

$$\text{minimize} \sum_{i\in K_1} g_i + \sum_{i\in K_2} g_i y_i + \sum_{i\in K_1\cup K_2} \sum_j a_{ij} x_{ij} \tag{43}$$

under the constraints (28) to (31), has the same form as the simple plant-location problem. Let

$$a'_{ij} = \begin{cases} a_{ij} + g_i/n_i & \forall i \in K_2 \\ a_{ij} & \forall i \in K_1 \end{cases} \tag{44}$$

where the n_i are given by (19) (and not by a similar formula for the a_{ij}'s because constraints (25) and (26) imply that the public from each population center must go to the nearest facility and not necessarily to the facility the proportional investment cost of which is lowest). Further, let

$$a'_{rj} = \min_{i\in K_1\cup K_2} a'_{ij} \tag{45}$$

$$a'_{sj} = \min_{i\in K_1\cup K_2,\, i\neq r} a'_{ij}; \tag{46}$$

a reasoning parallel to that of Efroymson and Ray shows that

$$g = \sum_{i\in K_1} g_i + \sum_j a'_{rj} \tag{47}$$

is a lower bound of the values of (43) on $I'(K_0, K_1, K_2)$.

Moreover, the penalties

$$h_i^0 = \sum_j \max(a'_{sj} - a'_{ij}, 0) \tag{48}$$

and

$$h_i^1 = g_i - \sum_{j\mid r\in K_1\cup\{i\}} \max(a'_{rj} - a_{ij}, 0) - \sum_{j\mid r\in K_2\setminus\{i\}} \max(a'_{sj} - a_{ij}, 0), \tag{49}$$

defined for all $i \in K_2$, are additive, and

$$g^1 = g + \sum_{i \in K_2} \min(h_i^0, h_i^1) \tag{50}$$

is a lower bound of the values of (43) on $I'(K_0, K_1, K_2)$.

Finally, let, for every $i \in K_2$,

$$e_i = |h_i^0 - h_i^1| \tag{51}$$

and

$$K_5 = \{i \in K_2 \mid h_i^0 \geq h_i^1\} \tag{52}$$

$$K_6 = \{i \in K_2 \mid h_i^1 > h_i^0\} \tag{53}$$

$$m_5 = |K_5| \tag{54}$$

$$m_6 = |K_6|\,; \tag{55}$$

e_k^0, $k = 1, 2, \ldots, m_5$ denote the e_i's the indices of which are in K_5, ranked in order of increasing value, and e_k^1, $k = 1, 2, \ldots, m_6$ denote the e_i's the indices of which are in K_6, also ranked in order of increasing value. Then

$$g^2 = \begin{cases} g^1 + \sum_{k=1}^{p-m_1-m_5} e_k^1 & \text{if} \quad m_1 + m_5 < p \\ g^1 & \text{if} \quad m_1 + m_5 = p \\ g^1 + \sum_{k=1}^{m_1+m_5-p} e_k^0 & \text{if} \quad m_1 + m_5 > p \end{cases} \tag{56}$$

is a lower bound of the values taken by (43) on $I'(K_0, K_1, K_2)$.

Another way of obtaining a lower bound of the values of (23) and of (43) (which generalizes some results of R. Zimmerman [15] and of K. Spielberg [12]) will also be used in the algorithm and will lead to efficient tests. Let us notice that, if no other facilities than those already established at a current iteration will be opened, the value of (23) will be

$$f^* = m_1 f + \sum_j c_{rj}^*. \tag{57}$$

If a facility is opened in the location $i \in K_2$ the decrease (or increase) in the value taken by (23) will be

$$t_i = f - \sum_j \max(c_{rj}^* - c_{ij}, 0). \tag{58}$$

Let t_k, $k = 1, 2, \ldots, m_2$ denote the t_i's ranked in order of increasing value.

Then clearly,

$$f^3 = f^* + \sum_{k=1}^{p-m_1} t_k \tag{59}$$

is a lower bound of the values taken by (23) on $I'(K_0, K_1, K_2)$.

Let us now consider the investment constraint; at a current iteration let $\tilde{a}_{rj}$ denote the proportional investment cost for service of the population of the jth center at the nearest already established facility:

$$\tilde{a}_{rj} = \min_{i \in K_1 \mid c_{ij} = c^*_{rj}} a_{ij}. \tag{60}$$

If no new facilities were established, total investment costs would be

$$g^* = \sum_{i \in K_1} g_i + \sum_j \tilde{a}_{rj}. \tag{61}$$

If a facility were opened in the location $i \in K_2$, the decrease (or increase) in total investment costs would be at most

$$l_i = g_i - \sum_{j \mid c_{ij} \leqslant c^*_{rj}} \max(\tilde{a}_{rj} - a_{ij}, 0); \tag{62}$$

let l_k, $k = 1, 2, \ldots, m_2$ denote the l_i's ranked in order of increasing value. Then clearly,

$$g^3 = g^* + \sum_{k=1}^{p-m_1} l_k \tag{63}$$

is a lower bound of the values taken by (43) on $I'(K_0, K_1, K_2)$.

4. Algorithm

A. Initialization

A.1. Set $p = 0$.

Set f_{opt}, value of the best solution obtained so far at a large value.

Set $y_i^{\text{opt}} = \varphi$ for $i = 1, 2, \ldots, m$ (best solution obtained so far).

A.2. Set $p = p + 1$. If $p > m$, stop.

Set $y_i = \varphi$ (not-yet-fixed boolean value) for $i = 1, 2, \ldots, m$. Set $M = \emptyset$ (vector in which will be recorded the indices of the variables used in the separations).

B. Tests

B.1. *First Direct Feasibility Test*

a) If $m_2 = p - m_1$ set $y_i = 1$, $\forall i \in K_2$, add all indices $i \in K_2$ to M by the right and underline them. Then go to B.10.

b) If $m_1 = p$ set $y_i = 0$, $\forall i \in K_2$, add all indices $i \in K_2$ to M by the right and underline them. Then go to B.10.

B.2. *First Direct Optimality Test*

Compute $\underline{f}$ by formula (32).
Compute $\forall i \in K_2$ the additive penalties p_i^0 and p_i^1 by formulas (34) and (35).
Compute $\underline{f}^1$ by formula (36).
Compute $\forall i \in K_2$ the d_i's by formula (37).
Rank the d_i's the indices of which are in K_3 (i.e. the d_i's such that $p_i^0 \geq p_i^1$) and the d_i's the indices of which are in K_4 (i.e. the d$_i$'s such that $p_i^1 > p_i^0$) in order of increasing value. Let d_k^0, $k = 1, 2, \ldots, m_3$ and d_k^1, $k = 1, 2, \ldots, m_4$ denote these d_i's after ranking; i_k^0, $k = 1, 2, \ldots, m_3$ and i_k^1, $k = 1, 2, \ldots, m_4$ the corresponding indices.
Compute $\underline{f}^2$ by formula (42).
If $\underline{f}^2 \geq f_{\text{opt}}$ go to D (backtracking).

B.3. *First Conditional Optimality Test*

a) First case: $m_1 + m_3 < p$.

a.1. Let $k^* = p - m_1 - m_3 + 1$.
Compute for $k = m_3, m_3 - 1, \ldots, 1$:

$$\underline{f}_k^2 = \underline{f}^2 + d_k^0 + d_{k^*}^1 .$$

If $\underline{f}_k^2 < f_{\text{opt}}$ go to a.2.
Otherwise fix $y_{i_k^0}$ at 1, add the index i_k^0 to M by the right and underline it.

a.2. Compute for $k = 1, 2, \ldots, k^* - 1$:

$$\underline{f}_k^2 = \underline{f}^2 + d_{k^*}^1 - d_k^1 .$$

If $\underline{f}_k^2 < f_{\text{opt}}$ go to a.3.
Otherwise fix $y_{i_k^1}$ at 1, add the index i_k^1 to M by the right and underline it.

a.3. Let $k^{**}=p-m_1-m_3$.
Compute for $k=m_4, m_4-1, \ldots, k^*$:

$$\underline{f}_k^2=\underline{f}^2+d_k^1-d_{k^{**}}^1.$$

If $\underline{f}_k^2<f_{\text{opt}}$ go to B.4.
Otherwise fix $y_{i_k^1}$ at 0, add the index i_k^1 to M by the right and underline it.

b) Second case: $m_1+m_3=p$.

b.1. Compute for $k=m_3, m_3-1, \ldots, 1$:

$$\underline{f}_k^2=\underline{f}^2+d_k^0+d_1^1.$$

If $\underline{f}_k^2<f_{\text{opt}}$ go to b.2.
Otherwise fix $y_{i_k^0}$ at 1, add the index i_k^0 to M by the right and underline it.

b.2. Compute for $k=m_4, m_4-1, \ldots, 1$:

$$\underline{f}_k^2=\underline{f}^2+d_k^1+d_1^0.$$

If $\underline{f}_k^2<f_{\text{opt}}$ go to B.4.
Otherwise fix $y_{i_k^1}$ at 0, add the index i_k^1 to M by the right and underline it.

c) Third case: $m_1+m_3>p$.

c.1. Let $k^*=m_1+m_3-p+1$.
Compute for $k=1, 2, \ldots, k^*-1$:

$$\underline{f}_k^2=\underline{f}^2+d_{k^*}^0-d_k^0.$$

If $\underline{f}_k^2<f_{\text{opt}}$ go to c.2.
Otherwise fix $y_{i_k^0}$ at 0, add the index i_k^0 to M by the right and underline it.

c.2. Let $k^{**}=m_1+m_3-p$.
Compute for $k=m_3, m_3-1, \ldots, k^*$:

$$\underline{f}_k^2=\underline{f}^2+d_k^0-d_{k^{**}}^0.$$

If $\underline{f}_k^2<f_{\text{opt}}$ go to c.3.
Otherwise fix $y_{i_k^0}$ at 1, add the index i_k^0 to M by the right and underline it.

c.3. Compute for $k=m_4, m_4-1, \ldots, 1$:

$$\underline{f}_k^2=\underline{f}^2+d_k^1+d_{k^*}^0.$$

If $\underline{f}_k^2<f_{\text{opt}}$ go to B.4.
Otherwise fix $y_{i_k^1}$ at 0, add the index i_k^1 to M by the right and underline it.

B.4. *Second Direct Feasibility Test*

Compute $\underline{g}$ by formula (47).
Compute $\forall i \in K_2$ the additive penalties h_i^0 and h_i^1 by formulas (48) and (49).
Compute $\underline{g}^1$ by formula (50).
Compute $\forall i \in K_2$ the e_i's by formula (51).

Rank the e_i's the indices of which are in K_5 (i.e. the e_i's such that $h_i^0 \geq h_i^1$) and the e_i's the indices of which are in K_6 (i.e. the e_i's such that $h_i^1 > h_i^0$) in order of increasing value. Let e_k^0, $k = 1, 2, \ldots, m_5$ and e_k^1, $k = 1, 2, \ldots, m_6$ denote these e_i's after ranking i_k^0, $k = 1, 2, \ldots, m_5$ and i_k^1, $k = 1, 2, \ldots, m_6$, the corresponding indices.

Compute $\underline{g}^2$ by formula (56).
If $\underline{g}^2 > b$ go to *D*.

B.5. *First Conditional Feasibility Test*

a) First case: $m_1 + m_5 < p$.

a.1. Let $k^* = p - m_1 - m_5 + 1$.
Compute for $k = m_5, m_5 - 1, \ldots, 1$:

$$\underline{g}_k^2 = \underline{g}^2 + e_k^0 + e_{k^*}^1 .$$

If $\underline{g}_k^2 \leq b$ go to a.2.
Otherwise fix $y_{i_k^0}$ at 1, add the index i_k^0 to M by the right and underline it.

a.2. Compute for $k = 1, 2, \ldots, k^* - 1$:

$$\underline{g}_k^2 = \underline{g}^2 + e_{k^*}^1 - e_k^1 .$$

If $\underline{g}_k^2 \leq b$ go to a.3.
Otherwise fix $y_{i_k^1}$ at 1, add the index i_k^1 to M by the right and underline it.

a.3. Let $k^{**} = p - m_1 - m_5$.
Compute for $k = m_6, m_6 - 1, \ldots, k^*$:

$$\underline{g}_k^2 = \underline{g}^2 + e_k^1 - e_{k^{**}}^1 .$$

If $\underline{g}_k^2 \leq b$ go to **B**.6.
Otherwise fix $y_{i_k^1}$ at 0, add the index i_k^1 to M by the right and underline it.

b) Second case: $m_1 + m_5 = p$.

b.1. Compute for $k = m_5, m_5 - 1, \ldots, 1$:

$$\underline{g}_k^2 = \underline{g}^2 + e_k^0 + e_1^1 .$$

If $\underline{g}_k^2 \leq b$ go to b.2.
Otherwise fix $y_{i_k^0}$ at 1, add the index i_k^0 to M by the right and underline it.

b.2. Compute for $k = m_6, m_6 - 1, \ldots, 1$:

$$\underline{g}_k^2 = \underline{g}^2 + e_k^1 + e_1^0.$$

If $\underline{g}_k^2 \leq b$ go to B.6.
Otherwise fix $y_{i_k^1}$ at 0, add the index i_k^1 to M by the right and underline it.

c) Third case: $m_1 + m_5 > p$.

c.1. Let $k^* = m_1 + m_5 - p + 1$.
Compute for $k = 1, 2, \ldots, k^* - 1$:

$$\underline{g}_k^2 = \underline{g}^2 + e_{k^*}^0 - e_k^0.$$

If $\underline{g}_k^2 \leq b$ go to c.2.
Otherwise fix $y_{i_k^0}$ at 0, add the index i_k^0 to M by the right and underline it.

c.2. Let $k^{**} = m_1 + m_5 - p$.
Compute for $k = m_5, m_5 - 1, \ldots, k^*$:

$$\underline{g}_k^2 = \underline{g}^2 + e_k^0 - e_{k^{**}}^0.$$

If $\underline{g}_k^2 \leq b$ go to c.3.
Otherwise fix $y_{i_k^0}$ at 1, add the index i_k^0 to M by the right and underline it.

c.3. Compute for $k = m_4, m_4 - 1, \ldots, 1$:

$$\underline{g}_k^2 = \underline{g}^2 + e_k^1 + e_{k^*}^0.$$

If $\underline{g}_k^2 \leq b$ go to B.6.
Otherwise fix $y_{i_k^1}$ at 0, add the index i_k^1 to M by the right and underline it.

B.6. *Second Direct Optimality Test*

Compute f^* by formula (57).

Compute $\forall i \in K_2$ the t_i's by formula (58).

Rank the t_i's in order of increasing value. Let t_k, $k = 1, 2, \ldots, m_2$ denote these t_i's after ranking and i_k, $k = 1, 2, \ldots, m_2$ the corresponding indices.

Compute $\underline{f}^3$ by formula (59).

If $\underline{f}^3 \geq f_{\text{opt}}$ go to *D*.

B.7. *Second Conditional Optimality Test*

a) Let $k^* = p - m_1 + 1$ and $k^{**} = p - m_1$.
Compute for $k = 1, 2, \ldots, k^{**}$:

$$\underline{f}_k^3 = \underline{f}^3 - t_k + t_{k^{**}}.$$

If $\underline{f}_k^3 < f_{\text{opt}}$ go to b).
Otherwise fix y_{i_k} at 1, add the index i_k to M by the right and underline it.

b) Compute for $k=m_2, m_2-1, \ldots, k^*$:

$$\underline{f}_k^3=\underline{f}^3+t_k-t_{k^*}.$$

If $\underline{f}_k^3<f_{\text{opt}}$ go to B.8.
Otherwise fix y_{i_k} at 0, add the index i_k to M by the right and underline it.

B.8. *Third Direct Feasibility Test*

Compute g^* by formula (61).
Compute $\forall i \in K_2$ the l_i's by formula (62).
Rank the l_i's in order of increasing value. Let l_k, $k=1,2,\ldots,m_2$ denote these l_i's after ranking and i_k, $k=1,2,\ldots,m_2$ the corresponding indices.
Compute g^3 by formula (63).
If $\underline{g}^3>b$ go to D.

B.9. *Second Conditional Feasibility Test*

a) Let $k^*=p-m_1+1$ and $k^{**}=p-m_1$.
Compute for $k=1,2,\ldots,k^{**}$:

$$\underline{g}_k^3=\underline{g}^3-l_k+l_{k^{**}}.$$

If $\underline{g}_k^3\leq b$ go to b).
Otherwise fix y_{i_k} at 1, add the index i_k to M by the right and underline it.

b) Compute for $k=m_2, m_2-1, \ldots, k^*$:

$$\underline{g}_k^3=\underline{g}^3+l_k-l_{k^*}.$$

If $\underline{g}_k^3\leq b$ go to B.10.
Otherwise fix y_{i_k} at 0, add the index i_k to M by the right and underline it.

B.10. *Resolution Test*

a) If at least one variable remains free ($K_2\neq\emptyset$) and at least one variable has been fixed at the value 0 or 1 during the last passage through the tests B.1. to B.9., return to test B.1.
b) If at least one variable remains free and no variable has been fixed at the value 0 or 1 during the last passage through the tests B.1. to B.9., go to C (choice step).
c) If no variable remains free, compute $\underline{f}$ by formula (32) and g^* by formula (61).
If $\underline{f}\geq f_{\text{opt}}$ or $g^*>b$ go to D.
Otherwise update f_{opt} by setting $f_{\text{opt}}=\underline{f}$, update Y^{opt} by setting $y_i^{\text{opt}}=y_i$ for $i=1,2,\ldots,m$. Then go to D.

C. *Choice*

Select among the free variables the variable y_{i*} such that

$$d_{i*} = \max_{i \in K_2} d_i .$$

If $p_{i*}^0 \geq p_{i*}^1$ set $y_{i*} = 1$, add the index i^* to M by the right and go to test B.1.
If $p_{i*}^1 > p_{i*}^0$ set $y_{i*} = 0$, add the index i^* to M by the right and go to test B.1.

D. *Backtracking*

Seek from right to left an index i^* not underlined in M.

If no such index exists, go back to A.2.

Otherwise erase all indices to the right of i^*, set the corresponding variables y_i at the value φ, set y_{i*} at the value 0 if it was at the value 1 and at the value 1 if it was at the value 0, underline i^* in M and go back to test B.1.

5. Computational Experience

The algorithm has been coded in FORTRAN EXTENDED and tested on a CDC 6400 computer. Thirteen test problems have been randomly generated with the following characteristics:

c_{ij}: between 10 and 100
a_{ij}: between 10 and 100
g_i : between 500 and 2000
b : 7000.

The value of the fixed costs f introduced in the objective function must be chosen in order to give the penalties p_i^0 and p_i^1 of formulas (34) and (35) approximatively the same values; f was fixed at 100. The number n of population centers was 100 for all problems. The number m of locations where central facilities could be established was between 10 and 22. Table 1 gives the costs of optimal solutions, the numbers of central facilities opened in these solutions, as well as computation times in seconds.

Table 2 analyzes the resolution of problem 13. Computation times are given for each value of p up to 10 as well as cumulated times, and the numbers of open facilities and costs of the best solution known so far.

In conclusion, the algorithm can solve CFLP's of realistic size. However, computation times are much larger than those for simple plant-location problems of the same size. Several reasons can be given for this. First, checking whether each partial solution satisfies constraints (2) and (3) is very time consuming. Second, because of the constraint on investment funds the

TABLE 1

Problem	n	m	f_{opt}	Number of facilities	Time
1	100	10	3700	2	54.374
2	100	10	3364	2	44.600
3	100	12	4847	1	31.266
4	100	12	3751	2	55.799
5	100	14	3821	2	91.525
6	100	14	3746	2	41.753
7	100	16	3709	2	162.853
8	100	16	4109	2	101.449
9	100	18	3695	2	232.439
10	100	18	3652	2	305.211
11	100	20	3686	2	449.889
12	100	20	3746	2	381.851
13	100	22	3117	3	694.322

TABLE 2

p	Time	Time cumulated	Cost	Number of facilities
1	12.946	12.946	4734	1
2	49.456	62.402	3778	2
3	180.882	243.284	3117	3
4	193.422	436.706	3117	3
5	156.507	593.213	3117	3
6	79.961	673.174	3117	3
7	1.923	675.097	3117	3
8	1.362	676.459	3117	3
9	1.353	677.812	3117	3
10	1.355	679.167	3117	3
11→22	15.155	694.322	3117	3

variables y_i may not be fixed at 0 (as in algorithms for the simple plant-location problem) when the maximum transportation cost reduction when opening a facility in location i is less than f. Third, because the public from each population center goes to the nearest facility (i.e. with smallest c_{ij}), and not to the facility with smallest a_{ij}, lower bounds on the values of the left-hand side of (2) will tend to be looser than those on the values of the objective function and the corresponding tests will be less efficient.

References

1. BALINSKI, M. L. On finding integer solutions to linear problems. *Mathematica*, Princeton, New Jersey, 1964.
2. EFROYMSON, M. A. and RAY, T. L. A branch-and-bound algorithm for plant location. *Operations Research*, **14**: 361–368, 1966.
3. GEOFFRION, A. Integer programming by implicit enumeration and Balas' method. *S.I.A.M. Review*, **9**: 178–190, 1967.
4. HAKIMI, S. L. Optimum distribution of switching centers in a communication network and some related graph theoretic problems. *Operations Research*, **13**: 462–475, 1965.
5. HANSEN, P. Pénalités additives pour les programmes en variables zéro-un. *Comptes-Rendus de l'Académie des Sciences de Paris*, **273**: 175–177, 1971.
6. HANSEN, P. Pénalités additives pour le problème de la localisation des entrepôts. *Comptes-Rendus de l'Académie des Sciences de Paris*, **273**: 252–253, 1971.
7. HANSEN, P. Two algorithms for the simple plant-location problem using additive penalties. Working paper. University of Brussels, 1972.
8. HANSEN, P. An algorithm for the weighted *p*-median of a graph. Working paper, University of Brussels, 1972.
9. MARANZANA, F. E. On the location of supply points to minimize transport costs. *Operational Research Quarterly*, **15**: 261–270, 1964.
10. REVELLE, C. and ROGESKI, P. Central facilities location under an investment constraint. *Geographical Analysis*, **2**: 343–360, 1970.
11. REVELLE, C. and SWAIN, R. Central facilities location. *Geographical Analysis*, **2**: 30–42, 1970.
12. SPIELBERG, K. Algorithm for the simple plant-location problem with some side conditions. *Operations Research*, **17**: 85–111, 1969.
13. SPIELBERG, K. Plant location with generalized search origin. *Management Science*, **16**: 165-178, 1969.
14. TEITZ, M. and BART, P. Heuristic methods for estimated generalized vertex median of a weighted graph. *Operations Research*, **16**: 955–962, 1968.
15. ZIMMERMAN, R. A branch-and-bound algorithm for depot location. *Metra*, **4**: 661-674, 1967.

CHAPTER 12

COMPUTER METHODS FOR SCHOOL SCHEDULES

M. ROUBENS

Summary

Two general methods are used when dealing with the computation of timetables: (1) graph partitioning to solve examination schedules; (2) allocation techniques to construct school timetables. A short description is given of the existing algorithms used to solve such problems. Two heuristics suited to the construction of school timetables are described.

The first heuristic proposes a guided walk through the availability matrix using the concept of float protection to provide the timetable matrix. Allocation of rooms can be performed without difficulty. The second heuristic is based on a modified version of the well-known allocation algorithm called Hungarian method. It can be used successfully for tight room restrictions. Moreover, binary availability is replaced by a utility concept.

1. Introduction

Timetable construction is a typical scheduling problem, i.e. arrangement of events (examinations, lectures) subject to constraints.

1.1. When constructing an examination timetable the following constraints generally have to be satisfied:

(1.1.1) no candidate is required to take two examinations at the same time;

(1.1.2) each room can only accommodate a certain number of seats;

(1.1.3) all candidates for a subject must be in the same room;

(1.1.4) candidates should have their examinations distributed so that they do not occur too closely together.

1.2. When assigning teachers to classes, one has to take into account:

(1.2.1) the number of hours a given teacher must meet a given class during one week;

(1.2.2) times when some class, teacher, or room is not available for teaching;

(1.2.3) each lecture is given by one teacher in one class;

(1.2.4) some lectures require special rooms (laboratory, gymnasium) and each room can only accommodate a certain number of seats.

2. Examination Schedules

The fundamental requirement is that no candidate is required to take two examinations at the same time. This restriction can be resumed in a *conflict matrix* $M = \{M(i,j)\}$ where

$$M(i,j) \begin{cases} = 1 \text{ if subjects } (i) \text{ and } (j) \text{ cannot take place concurrently} \\ = 0 \text{ otherwise.} \end{cases}$$

Scheduling the examinations subject to constraint (1.1.1) is equivalent to coloring a graph, whose boolean incidence matrix is M, such that no two adjacent vertices have the same color.

The method widely used for this type of problem is to arrange the vertices in decreasing order of degree (number of edges having the vertex as their endpoint). Welsh and Powell [13] used this procedure. Wood [15] modified the preceding algorithm by considering a similitude matrix associated with the conflict matrix. No assurance is given that these heuristics lead to the minimum number of colors. The chromatic number is hard to calculate but an upper bound [13] is easily determined. Anthonisse [2] proposed a Branch-and-Bound method to solve this problem.

The constraints (1.1.2) to (1.1.4) have been introduced in the procedures developed by Herz [8], Peck and Williams [12], and Wood [14].

3. School Timetables

Given a set of teachers $T(i)$, a set of classes $C(j)$, a set of rooms $R(l)$, the problem is to allocate hours $H(k)$ to classes, teachers to classes, rooms to classes, such that conditions (1.2.1) to (1.2.4) are satisfied.

3.1. If allocation of rooms is not considered, the conditions (1.2.1) to (1.2.3) can be resumed into matrices G, $ATCH$:

G, *the class requirements matrix* with integer elements $G(i, j)$ giving the number of times teacher $T(i)$ must meet class $C(j)$ in a week;

$ATCH$, *the availability matrix* with elements $ATCH(i, j, k)$ which are 1 or 0 according as $T(i)$ is able or not, to meet $C(j)$ at hour $H(k)$.

Csima and Gotlieb [3] presented a *tight search algorithm* to reduce the availability matrix to a timetable, i.e. an array with elements satisfying

$$\begin{cases} \Sigma_k \ ATCH(i, j, k) = G(i, j) \\ \Sigma_i \ ATCH(i, j, k) \leq 1 \\ \Sigma_j \ ATCH(i, j, k) \leq 1 \,. \end{cases}$$

Devices to improve the efficiency of the procedure were presented by Lions [10]. Computation times remain unsatisfactory, however. Moreover, for theoretical reasons, it is desirable to complete the availability matrix by adding dummy teachers, classes or hours to obtain a latin square timetable.

Almond [1] proposed a *random walk* from 1 to 1 in the availability matrix to construct the timetable without considering the feasibility of the procedure.

We propose in section 4—see also [4, 5]—a modified version of Almond's procedure called *float protection algorithm* which guides the walk through the availability matrix.

3.2. Slight modifications of the methods presented in 3.1 permit the allocation of rooms. Yule [16] introduced such modifications into the procedure of Almond by proposing an alternative method of information storage. Etienne [4] took up the same idea. Lejeune [9] developed a procedure whose input takes explicitly into account room accommodation and teacher preference for some periods of the week. This solution is described in section 5. Recently Gruyer, Pandolfi and Slabodsky [7] proposed a Branch-and-Bound method to solve the problem.

3.3. There remains to find N. and S. conditions to ensure the feasibility of any allocation procedure. Gotlieb [6] presented feasibility conditions—called

Hall conditions— and conjectured that his method would always be sufficient but Lions [11] refuted this argument with a counterexample.

4. The Float Protection Algorithm

Let us again consider the sets T, C, H and the matrices G and $ATCH$. The information contained in the 3-dimensional matrix $ATCH$ can be resumed to compose two 2-dimensional boolean arrays.

ATH, the teacher-hour availability array
ACH, the class-hour availability array.

The connection between these arrays and $ATCH$ is given by

$$ATH(i,k) = [\Sigma_j\, ATCH(i,j,k)]_{\text{mod. }1}$$
$$ACH(j,k) = [\Sigma_i\, ATCH(i,j,k)]_{\text{mod. }1}$$

$$ATH(i,k)\text{—resp. } ACH(j,k) \begin{cases} = 1 \text{ if teacher } T(i)\text{—resp. class } C(j)\text{—is available at hour } H(k) \\ = 0 \text{ otherwise.} \end{cases}$$

Finally, the following floats are introduced:

FTC, the teacher-class float matrix
FT, the teacher float vector
FC, the class float vector.

These concepts are defined as:

$$FTC(i,j) = \begin{cases} \Sigma_k\, ATH(i,k)\, ACH(j,k) - G(i,j), & \text{if } G(i,j) \neq 0 \\ \text{undefined} & \text{if } G(i,j) = 0 \end{cases}$$
$$FT(i) = \Sigma_k\, ATH(i,k) - \Sigma_j\, G(i,j)$$
$$FC(j) = \Sigma_k\, ACH(j,k) - \Sigma_i\, G(i,j).$$

To illustrate, suppose a class requirements matrix and availability arrays as shown in the following tables (4 teachers, 3 classes, 8 hours a week).

$G(i,j)$:

	$C(1)$	$C(2)$	$C(3)$
$T(1)$	1	0	3
$T(2)$	0	3	2
$T(3)$	2	0	0
$T(4)$	2	2	2

$$ATH(i,k):\quad \begin{array}{c} \\ T(1) \\ T(2) \\ T(3) \\ T(4) \end{array} \begin{array}{cccccccc} H(1) & H(2) & H(3) & H(4) & H(5) & H(6) & H(7) & H(8) \\ \end{array}$$

$$ATH(i,k):\quad \begin{array}{c|cccccccc} & H(1) & H(2) & H(3) & H(4) & H(5) & H(6) & H(7) & H(8) \\ \hline T(1) & 1 & 1 & 1 & 1 & 0 & 0 & 0 & 0 \\ T(2) & 1 & 1 & 1 & 0 & 1 & 1 & 1 & 1 \\ T(3) & 0 & 0 & 0 & 0 & 0 & 0 & 1 & 1 \\ T(4) & 1 & 1 & 1 & 0 & 1 & 1 & 1 & 1 \end{array}$$

$$ACH(j,k):\quad \begin{array}{c|cccccccc} & H(1) & H(2) & H(3) & H(4) & H(5) & H(6) & H(7) & H(8) \\ \hline C(1) & 1 & 1 & 0 & 0 & 1 & 1 & 1 & 1 \\ C(2) & 1 & 1 & 1 & 0 & 0 & 0 & 1 & 1 \\ C(3) & 1 & 1 & 1 & 1 & 1 & 1 & 1 & 1 \end{array}.$$

The input just given permits to obtain

$$FTC(i,j):\quad \begin{array}{c|ccc} & C(1) & C(2) & C(3) \\ \hline T(1) & 1 & \times & 1 \\ T(2) & \times & 2 & 5 \\ T(3) & 0 & \times & \times \\ T(4) & 4 & 3 & 5 \end{array} \qquad FT(i):\quad \begin{bmatrix} 0 \\ 2 \\ 0 \\ 1 \end{bmatrix}$$

$$FC(j):\quad (\;1 \quad 0 \quad 1\;).$$

At each step of the procedure one teacher is allocated to one class at some hour. The matrices G, ATH, ACH and the different floats are updated. Instead of allocating a triple (T, C, H) in a random way as required in Almond's procedure, we studied in [5] the impact of an allocation on the floats.

As can easily be seen, these floats are nonnegative if the partial solution is feasible and they never increase their value during the procedure. The heuristic works so as to *protect the critical floats*.

The procedure involves a search for the triple $(T(t), C(c), H(h))$ according to the following hierarchical criteria:

— CRITERION 1: affect, by priority, teacher $T(t)$ which is the only candidate to class $C(c)$ at hour $H(h)$, $h = 1, 2, \ldots$.

— CRITERION 2: affect then class $C(c)$ which is the only candidate to teacher $T(t)$ at hour $H(R)$, $h=1, 2, \ldots$.

— CRITERION 3: affect $T(t)$, $C(c)$, $H(h)$ according to $FTC(t, c)=0$, $FT(t)=0, FC(c)=0$.

— CRITERION 4: protect two critical floats.

— CRITERION 5: protect one critical float.

The justification of rules 1 and 2 is due to the fact that the allocation of $T(t)$, $C(c)$, $H(h)$ affects the floats as

$FT(t)$ is:
- — decreasing by b units IF $G(t, c)=0$ after updating and IF $C(c)$ is the only candidate to $T(t)$ for b hours different from $H(h)$;
- — nondecreasing, otherwise.

$FC(c)$ is:
- — decreasing by b units IF $G(t, c)=0$ after updating and IF $T(t)$ is the only candidate to $C(c)$ for b hours different from $H(h)$;
- — nondecreasing, otherwise.

The other floats are decreasing by 1 or 0 unit.

In the example, one can find the following allocations:

CRITERION 1, first pass:

$H(4)\ C(3)\ T(1)$ $H(5)\ C(1)\ T(4)$ $H(5)\ C(3)\ T(2)$
$H(6)\ C(1)\ T(4)$ $H(6)\ C(3)\ T(2)$ $H(7)\ C(1)\ T(3)$
$H(7)\ C(3)\ T(4)$ $H(8)\ C(1)\ T(3)$ $H(8)\ C(3)\ T(4)$

second pass:

$H(1)\ C(1)\ T(1)$ $H(2)\ C(3)\ T(1)$ $H(3)\ C(3)\ T(1)$
$H(7)\ C(2)\ T(2)$ $H(8)\ C(2)\ T(2)$

CRITERION 2, first pass:

$H(1)\ T(2)\ C(2)$ $H(2)\ T(4)\ C(2)$ $H(3)\ T(4)\ C(2)$

END.

The resulting timetable is

	$H(1)$	$H(2)$	$H(3)$	$H(4)$	$H(5)$	$H(6)$	$H(7)$	$H(8)$
$C(1)$	$T(1)$				$T(4)$	$T(4)$	$T(3)$	$T(3)$
$C(2)$	$T(2)$	$T(4)$	$T(4)$				$T(2)$	$T(2)$
$C(3)$		$T(1)$	$T(1)$	$T(1)$	$T(2)$	$T(2)$	$T(4)$	$T(4)$

5. Allocation Algorithm of Lejeune

Let us now consider:

— the sets T, C, H, R;

— the class requirements matrix G;

— the *preference matrix* P whose elements $P(i, k)$ indicate the preference of teacher $T(i)$ for hour $H(k)$: 1 indicates the highest preference, 9 indicates the lowest preference. Any other preference lies on the ordinal scale 2, 3, …, 8. Moreover 999 indicates that no teacher is available.

— the *effectiveness matrix* $E = E(m, n) = E\{(i, j), (k, l)\}$; $m = 1, \ldots, \Sigma_i \Sigma_j G(i, j)$; $m = 1, \ldots$; total number of hours $\times$ number of rooms.

$$E\{(i,j), (k,l)\} = P(i,k) \cdot \frac{\text{number of seats in room } R(l)}{\text{number of students in class } C(j)}$$

$$E\{(i,j),(k,l)\} \begin{cases} = 1 & \text{IF teacher } T(i) \text{ has highest preference for hour } H(k) \text{ and IF room } R(l) \text{ is perfectly accommodated with class } C(j) \\ > 1 & \text{otherwise.} \end{cases}$$

Lejeune proposes to solve the allocation problem:

$$\min \Sigma_m \Sigma_n \, x(m, n) \, E(m, n)$$

subject to

(5.1) $\Sigma_m \, x(m, n) \leq 1$

(5.2) $\Sigma_n \, x(m, n) \leq 1$

(5.3) $\Sigma_m \Sigma_n \, x(m,n) = \Sigma_i \Sigma_j \, G(i,j)$

(5.4) $x(m,n) = 0, 1$

(5.5) $x(m,n) = 1 \rightarrow x(m',n') = 0 \quad \text{IF} \quad m = (i,j);\ n = (k,l)$
$$m' = (i',j');\ n' = (k,l'),\ l' \neq l.$$

If condition (5.5) is relaxed, the above integer program is a classical assignment program. To meet condition (5.5) Lejeune [9] developed a procedure which follows the steps of the Hungarian method of Kuhn.

References[1]

1. ALMOND, M. An algorithm for constructing university timetables. *C.J.*, **8**: 331–340, 1966.
2. ANTHONISSE, J. M. Optimal partitioning, with an application to the construction of timetables. European Meeting on Statistics, Economics and Management Science, Amsterdam, 1968.
3. CSIMA, J. and GOTLIEB, C. C. Tests on computer method for constructing school timetables. *C.A.C.M.*, **7**: 160–163, 1964.
4. ETIENNE, J. M. Etablissement des horaires sur calculateur numérique. Application aux horaires de cours et d'examens d'université. Thèse, Faculté polytechnique de Mons, 1971.
5. ETIENNE, J. M. and ROUBENS, M. Formulation du problème des horaires pour un traitement sur ordinateur. Document RO. 7202, Faculté polytechnique de Mons, 1972.
6. GOTLIEB, C. C. The construction of class-teacher timetables. *Proc. IFIP Congress* 1962, Munich. North-Holland Publishing Cy, Amsterdam, 1963.
7. GRUYER, A., PANDOLFI, A. and SLABODSKY, G. Application de la théorie des graphes aux problèmes d'emploi du temps. *Thèse de doctorat* 3me cycle, Université de Paris, 1972.
8. HERZ, J. C. Quelques considérations sur les problèmes d'emploi du temps. *Revue Française d'Informatique et de Recherche Opérationnelle*, No.38: 85–91, 1966.
9. LEJEUNE, J. Contribution à l'élaboration d'un horaire de faculté par une approche formalisée. Thèse, Université de Liège, 1972.
10. LIONS, J. Matrix reduction using the Hungarian method for the generation of timetables. *C.A.C.M.*, **9**: 349–354, 1966.
11. LIONS, J. A counterexample for Gotlieb's method for the construction of school timetables. *C.A.C.M.*, **9**: 697–698, 1966.
12. PECK, J. E. L. and WILLIAMS, M. R. Examination algorithm, Algol algorithm 286. *C.A.C.M.*, **9**: 133–134, 1966.

---

[1] C.J. and C.A.C.M. abbreviate *Computer Journal* and *Communications of the Association of Computing Machinary*, respectively.

13. WELSH, D. J. A. and POWELL, M. B. An upper bound for the chromatic number of a graph and its application to timetabling problems. *C.J.*, **10**: 85–86, 1967.
14. WOOD, D. C. A system for computing university examination timetables. *C.J.*, **11**: 41–47, 1968.
15. WOOD, D. C. A technique for coloring a graph applicable to a large scale timetabling problem. *C.J.*, **12**: 317–319, 1969.
16. YULE, A. P. Extensions to the heuristic algorithm for university timetables. *C.J.*, **10**: 360–364, 1968.

CHAPTER 13

PROBABILISTIC MODELS FOR A DATA MULTIPLEXER[1]

MICHAEL RUBINOVITCH

Summary

A probabilistic model for data multiplexers with two input lines and one output line is formulated. It assumes that each input line alternates between activity and idleness and that both active and idle periods follow (possibly different) exponential distributions. The results include the distribution of the active periods on the output line and explicit expressions for the limiting distribution of the amount of data in storage.

1. Introduction

As the use of remote computers in business and science became more widespread, the utilization of buffers in data communication networks has rapidly grown. In contrast to human communications (such as telephone conversations), data communications often have the feature that immediate transmission is not essential and small delays could be tolerated. This makes it possible to introduce buffers, or data storage facilities, at various locations in a network for greater efficiency.

Suppose, for example, that a packet of data is to be transmitted from terminal A to terminal B in a network. If the transmission line leading from A to B is busy the packet could be stored in a buffer until the line becomes free. This would cut on the waiting time for transmission in A, and may help achieve a better utilization of the transmission line.

This paper considers a stochastic model for the analysis of a one-buffer data transmission system. The system, called a *multiplexer*, consists of a single buffer, two input lines leading to it, and one line, the main line, leading

[1] This research was carried out while the author was visiting with Bell Laboratories in Holmdel, N. J. I thank my colleagues at Bell Laboratories, Jerry Hayes and David Sherman, for having brought this problem to my attention; and Shlomo Halfin and Moshe Segal for stimulating discussions during the course of the investigation.

out. Such multiplexers are widely used in various communication networks. As a particular example, one may think of the main line as being connected to the central processing unit of a digital computer, and of the input lines as connected to remote teletype machines.

We shall assume here that each of the input lines is active during intervals of time of random length which are separated by intervals of idleness whose length is also random.

In this paper we treat the case when idle and active periods are exponentially distributed and derive the distribution function (d.f.) of the active periods on the main line as well as the limiting distribution of the content of the buffer.

2. Description of the Model

Consider a system consisting of a buffer, two *input lines*, line 1 and line 2, leading to it and one line, the *main* line, leading out.

Each of the input lines may be active or idle at any time, and when active it transmits data at a constant rate, say one symbol (bit) per time unit. The lengths of these active and idle periods are all mutually independent positive random variables.

The buffer is a device of infinite storage capacity which has the ability to accept data from several lines simultaneously, store it if necessary, and then transmit it to the main line. We shall not be concerned here with the logic of its operation nor with problems of data storage allocation in its memory (see Gaver and Lewis [1]). For our purpose the buffer is a "black box" which operates as follows.

Suppose that initially the input lines are idle and the buffer is empty. If, then, one of the lines becomes active the flow of data is transmitted to the main line without delay and the buffer acts as a mere connection. If, on the other hand, both input lines are active, then the buffer is being filled at a net rate of one bit per time unit and the main line is active. When the input lines are idle and the buffer is not empty, data are transmitted through the main line and the amount of data in storage decreases at rate 1.

We shall study this multiplexer buffer system under the following assumptions. Idle periods are exponentially distributed with mean λ_1^{-1} on line 1 and λ_2^{-1} on line 2. The lengths of the active periods are also exponential with mean μ_1^{-1} and μ_2^{-1} on line 1 and line 2, respectively. Here it is apparent that all active and idle periods on the main line are independent random variables and that the latter follow an exponential d.f. with mean $(\lambda_1+\lambda_2)^{-1}$. It, therefore, remains to derive the d.f. of the active period to completely characterize the traffic pattern on the main line.

3. The Active Period on the Main Line

In order to describe the state of our system we find it convenient to define a (right continuous) stochastic process $\{J(t),\ Z(t);\ t \geq 0\}$ as follows. Its first component $J(t)=0$ if both input lines are idle at time t, $J(t)=i\ (i=1, 2)$ if line number i is active and the other line is idle at time t, and $J(t)=3$ if both lines are active at time t. The second component $Z(t)$ denotes the amount of data in storage at time t. It is clear that $\{J(t),\ Z(t);\ t \geq 0\}$ is a Markov process.

Let

$$T = \inf\{t : Z(t) = J(t) = 0;\ \ Z(0) = J(0-) = 0;\ 1 \leq J(0) \leq 2\} \tag{1}$$

be the length of a typical active period on the main line and

$$\eta(\theta) = E\,\mathrm{e}^{-\theta T} \tag{2}$$

its Laplace-Stieltjes Transform (L.S.T.). Also, for $n=0, 1, 2, 3$ and $x \geq 0$ let

$$T_n(x) = \inf\{t : Z(t) = J(t) = 0;\ \ Z(0) = x,\ J(0) = n\} \tag{3}$$

and

$$\eta_n(\theta, x) = E\,\mathrm{e}^{-\theta T_n(x)}\,. \tag{4}$$

Thus, $T_n(x)$ is the remaining busy period on the main line when the system is in state i and the content of the buffer is x.

Suppose that initially the process $\{J(t),\ Z(t);\ t \geq 0\}$ is in state $(0, x)$. There are then three mutually exclusive cases : (i) The two idle input lines remain idle for an additional x time units and then the active period on the output line terminates; (ii) at some time $\tau < x$ line 1 becomes active while line 2 is still idle, in this case $T_0(x) = \tau + T_1(x-\tau)$; and (iii) at some time $\tau < x$ line 2 becomes active while line 1 is still idle and $T_0(x) = \tau + T_2(x-\tau)$. Hence with $\lambda = \lambda_1 + \lambda_2$ and $\mu = \mu_1 + \mu_2$ we may write

$$\begin{aligned} \eta_0(\theta, x) = \mathrm{e}^{-(\lambda+\theta)} + \int_0^x \lambda_1\,\mathrm{e}^{-(\lambda+\theta)\tau}\eta_1(\theta, x-\tau)\,\mathrm{d}\tau\, + \\ + \int_0^x \lambda_2\,\mathrm{e}^{-(\lambda+\theta)\tau}\eta_2(\theta, x-\tau)\,\mathrm{d}\tau\,. \end{aligned} \tag{5}$$

To simplify the last expression we consider the situation when $J(0)=1$, $Z(0)=x$ and let

(6) $$t_0 = \inf\{t : Z(t) < x\}.$$

Clearly $0 < t_0 < T_1(x)$ and it is not hard to see that t_0 and $T_1(x) - t_0$ are independent random variables and t_0 follows the same distribution as T while $T_1(x) - t_0$ is distributed like $T_0(x)$. Therefore $\eta_1(\theta, x) = \eta_1(\theta, 0)\, \eta_0(\theta, x)$ and similarly

(7) $$\eta_i(\theta, x) = \eta_i(\theta)\, \eta_0(\theta, x) \qquad (i = 1, 2, 3)$$

where $\eta_i(\theta) = \eta_i(\theta . 0)$.
This together with

(8) $$\eta(\theta) = \frac{\lambda_1}{\lambda} \eta_1(\theta) + \frac{\lambda_2}{\lambda} \eta(\theta)$$

is substituted into (5) to yield

$$\eta_0(\theta, x) = e^{-(\lambda+\theta)x} + \eta(\theta) \int_0^x e^{-(\lambda+\theta)\tau} \eta_0(\theta, x-\tau)\, d\tau,$$

which for fixed θ is a renewal equation for $\eta_0(\theta, x)$ with the boundary condition $\eta_0(\theta, 0) = 1$. The solution of this equation may be derived by standard procedures; we avoid the details and give the final result :

$$\eta_0(\theta, x) = e^{-[\lambda+\theta-\lambda\eta(\theta)]x},$$

and so

$$\eta_i(\theta, x) = \eta_i(\theta)\, e^{-[\lambda+\theta-\lambda\eta(\theta)]x}.$$

There are now four unknown functions $\eta(\theta)$, $\eta_1(\theta)$, $\eta_2(\theta)$, $\eta_3(\theta)$ and we proceed to derive more functional equations for their solution. These may be derived according to arguments similar to that used in the derivation of (5). For example, for $\eta_1(\theta, x)$ we have

$$\eta_1(\theta, x) = \int_0^\infty \mu_1 e^{-(\mu_1+\lambda_2+\theta)\tau} \eta_0(\theta, x)\, d\tau + \int_0^\infty \lambda_2 e^{-(\lambda_2+\mu_1+\theta)\tau} \eta_3(\theta, x)\, d\tau$$

$$= \frac{\mu_1}{\mu_1+\lambda_2+\theta} \eta_0(\theta, x) + \frac{\lambda_2}{\mu_1+\lambda_2+\theta} \eta_3(\theta, x),$$

which for $x=0$ reduces to

$$\eta_1(\theta)=\frac{\mu_1}{\mu_1+\lambda_2+\theta}\eta_0(\theta)+\frac{\lambda_2}{\mu_1+\lambda_2+\theta}\eta_3(\theta).$$

In the same way we can derive equations for η_2 and η_3 which together with (8) yield the following four equations for the four unknown L.S.T.:

$$\eta_1(\theta)=\frac{\mu_1}{\mu_1+\lambda_2+\theta}+\frac{\lambda_2}{\mu_1+\lambda_2+\theta}\eta_3(\theta) \tag{9}$$

$$\eta_2(\theta)=\frac{\mu_2}{\mu_2+\lambda_1+\theta}+\frac{\lambda_1}{\mu_2+\lambda_1+\theta}\eta_3(\theta) \tag{10}$$

$$\eta_3(\theta)=\frac{\mu_1\eta_2(\theta)+\mu_2\eta_1(\theta)}{\mu+\lambda+2\theta-\lambda\eta(\theta)} \tag{11}$$

$$\lambda\eta(\theta)=\lambda_1\eta_1(\theta)+\lambda_2\eta_2(\theta). \tag{12}$$

To solve for η we first substitute for η_2 and η_3 in (11) and (12) according to (9) and (10), respectively, and get

$$\eta_3(\theta)=\frac{\mu_1\mu_2[(\mu_1+\lambda_2+\theta)^{-1}+(\mu_2+\lambda_1+\theta)^{-1}]}{\mu+\lambda+2\theta-\lambda\eta(\theta)-\lambda_1\mu_1(\mu_2+\lambda_1+\theta)^{-1}-\lambda_2\mu_2(\mu_1+\lambda_2+\theta)^{-1}} \tag{13}$$

and

$$\eta_3(\theta)=\frac{\lambda\eta(\theta)-\lambda_1\mu_1(\mu_1+\lambda_2+\theta)^{-1}-\lambda_2\mu_2(\mu_2+\lambda_1+\theta)^{-1}}{\lambda_1\lambda_2[(\mu_1+\lambda_2+\theta)^{-1}+(\mu_2+\lambda_1+\theta)^{-1}]}. \tag{14}$$

Equating the right-hand sides of these equations we obtain a quadratic equation for η:

$$\lambda^2\eta(\theta)-C\eta(\theta)+D=0, \tag{15}$$

where

$$C=\lambda+\mu+2\theta+(\lambda_1\mu_1-\lambda_2\mu_2)\,[(\mu_1+\lambda_2+\theta)^{-1}-(\mu_2+\lambda_1+\theta)^{-1}]$$

$$\begin{aligned}D=\;&C[\lambda_1\mu_1(\mu_1+\lambda_2+\theta)^{-1}+\mu_2\lambda_2(\mu_2+\lambda_1+\theta)^{-1}]\\&-[\lambda_1\mu_1(\mu_1+\lambda_2+\theta)+\mu_2\lambda_2(\mu_2+\lambda_1+\theta)]^2\\&+4\lambda_1\lambda_2\mu_1\mu_2[(\mu_1+\lambda_2+\theta)^{-1}+(\mu_2+\lambda_1+\theta)^{-1}]^2.\end{aligned}$$

There are two solutions to this equation but one of them becomes unbounded as $\theta \to \infty$, and since η is an L.S.T. it must be ruled out. We are therefore left with only one solution, namely

$$\eta(\theta) = (2\lambda)^{-1}\left[\lambda+\mu+2\theta+\frac{(\lambda_1\mu_1-\lambda_2\mu_2)(\mu_2-\mu_1+\lambda_1-\lambda_2)}{(\mu_1+\lambda_2+\theta)(\mu_2+\lambda_1+\theta)}-\Delta\right] \tag{16}$$

where

$$\Delta = \frac{\lambda+\mu+2\theta}{(\mu_1+\lambda_2+\theta)(\mu_2+\lambda_1+\theta)}\{[(\mu_1+\lambda_2+\theta)(\mu_2+\lambda_1+\theta)-\lambda_1\mu_1-\lambda_2\mu_2]^2 - -4\lambda_1\lambda_2\mu_1\mu_2\}^{\frac{1}{2}}.$$

We have thus obtained the L.S.T. η of the active period on the main line.

Letting $\theta \to 0$ in (16) we find that

$$\Pr\{T<\infty\} = \lim_{\theta\to 0}\eta(\theta) = \begin{cases} 1 & \text{if} \quad \rho\leq 1 \\ \beta & \text{if} \quad \rho>1 \end{cases} \tag{17}$$

where

$$\rho = \frac{\lambda_1\lambda_2}{\mu_1\mu_2} \tag{18}$$

and

$$\beta = \frac{\mu_2(\mu_1+\lambda_2)^2+\mu_1(\mu_2+\lambda_1)^2}{\lambda(\mu_1+\lambda_2)(\mu_2+\lambda_1)}. \tag{19}$$

This can also be written as

$$\beta = \frac{2\lambda+\mu+\rho[\mu_2\lambda_2/\lambda_1+\mu_1\lambda_1/\lambda_2]}{\rho[2\lambda+\mu+\mu_2\lambda_2/\lambda_1+\mu_1\lambda_1/\lambda_2]}$$

which shows that $\beta<1$ when $\rho>1$ and $\beta=1$ when $\rho=1$.

When $\rho<1$ one can also obtain the expected value of the active period

$$\begin{aligned} E(T) &= \lim_{\theta\to 0+}\eta'(\theta) \\ &= \frac{\rho(\lambda+\mu)}{\lambda(1-\rho)(\mu_1+\lambda_2)(\mu_2+\lambda_1)}+\frac{\lambda_1}{\lambda(\mu_1+\lambda_2)}+\frac{\lambda_2}{\lambda(\mu_2+\lambda_1)} \end{aligned} \tag{20}$$

and $ET<\infty$.

Let us now examine the special case when the inputs on the two lines are the same, that is $\lambda_1 = \lambda_2 = \hat{\lambda}$, say, and $\mu_1 = \mu_2 = \hat{\mu}$. From (16) we have

$$\eta(\theta) = \frac{\hat{\lambda}+\hat{\mu}+\theta-\sqrt{(\hat{\lambda}+\hat{\mu}+\theta)^2-4\hat{\lambda}\hat{\mu}}}{2\hat{\lambda}}. \tag{21}$$

This is the L.S.T. of the busy period in an $M/M/1$ queue with input parameter $\hat{\lambda}$ and service rate $\hat{\mu}$. We also note that (21) may be inverted to obtain an explicit expression for the corresponding d.f. (see [2]).

4. Limiting Distributions

In this section we derive the limiting distribution of the process $\{Z(t), J(t)\}$ as defined in Section 3.

Let

$$F_i(x, t) = \Pr\{Z(t) \le x; \quad J(t) = i, Z(0) = J(0) = 0\} \tag{22}$$

for $x \ge 0$, $t \ge 0$ and $i = 0, 1, 2, 3$. It is easy to see that instants at which active periods on the main line terminate form a sequence of regeneration points for the $\{Z(t), J(t)\}$ process, and that times between successive points are independent identically distributed random variables. These random variables have absolutely continuous distribution functions and their means are finite when $\rho < 1$ (20). Thus it follows from renewal theory (Smith [3]) that when $\rho < 1$ (an assumption we shall make in the remainder of this section), $\{Z(t), J(t)\}$ possesses a proper limiting d.f. as $t \to \infty$. In other words

$$F_i(x) = \lim_{t\to\infty} F_i(x, t) \qquad (x \ge 0) \tag{23}$$

exist and if we let

$$F(x) = \sum_{i=0}^{3} F_i(x), \tag{24}$$

then $F(\infty) = 1$.

In the derivation of these functions we follow standard procedures. Writing first the forward Kolmogorov equations of the Markov process $\{Z(t), J(t)\}$ we then pass to the limit as $t \to \infty$. Explicit solutions to the resulting equations will then be found in a straightforward way. This is due to the fact that when $J(t) = 1$ or $J(t) = 2$ sample functions of $Z(t)$ are flat and the derivatives with respect to x of $F_1(x, t)$ and $F_2(x, t)$ vanish.

The forward Kolmogorov equations of our process are derived in the usual manner. They are:

$$\frac{\partial F_0(x,t)}{\partial t} - \frac{\partial F_0(x,t)}{\partial x} = -\lambda F_0(x,t) + \mu_1 F_1(x,t) + \mu_2 F_2(x,t) \tag{25}$$

$$\frac{\partial F_1(x,t)}{\partial t} = -(\mu_1+\lambda_2) F_1(x,t) + \mu_2 F_3(x,t) + \lambda_1 F_0(x,t) \tag{26}$$

$$\frac{\partial F_2(x,t)}{\partial t} = -(\mu_2+\lambda_1) F_2(x,t) + \mu_1 F_3(x,t) + \lambda_2 F_0(x,t) \tag{27}$$

$$\frac{\partial F_3(x,t)}{\partial t} + \frac{\partial F_3(x,t)}{\partial x} = -\mu F_3(x,t) + \lambda_1 F_2(x,t) + \lambda_2 F_1(x,t). \tag{28}$$

The limiting distributions $F_i(x)$ must therefore satisfy the following equations:

$$-F_0'(x) = -\lambda F_0(x) + \mu_1 F_1(x) + \mu_2 F_2(x) \tag{29}$$

$$0 = -(\mu_1+\lambda_2) F_1(x) + \mu_2 F_3(x) + \lambda_1 F_0(x) \tag{30}$$

$$0 = -(\mu_2+\lambda_1) F_2(x) + \mu_1 F_3(x) + \lambda_2 F_0(x) \tag{31}$$

$$F_3'(x) = -\mu F_3(x) + \lambda_1 F_2(x) + \lambda_2 F_1(x). \tag{32}$$

Substituting for $F_1(x)$ and $F_2(x)$ in (29) and (32) according to (30) and (31) we find that

$$F_0'(x) = F_3'(x) = \frac{F_0(x)\lambda_1\lambda_2(\lambda+\mu)}{(\mu_1+\lambda_2)(\mu_2+\lambda_1)} - \frac{F_3(x)\mu_1\mu_2(\lambda+\mu)}{(\mu_1+\lambda_2)(\mu_2+\lambda_1)}. \tag{33}$$

It follows that

$$F_0(x) = F_3(x) + F_0(0) \tag{34}$$

and hence

$$F_0(x) = \frac{F_0(0)}{1-\rho}[1-\rho e^{-\alpha x}] \tag{35}$$

$$F_3(x) = \frac{\rho F_0(0)}{1-\rho}[1-e^{-\alpha x}] \tag{36}$$

where

$$\alpha = \frac{(\lambda+\mu)(\mu_1\mu_2-\lambda_1\lambda_2)}{(\mu_1+\lambda_2)(\mu_2+\lambda_1)}. \tag{37}$$

Now we can use (35) and (36) in (30) and (31) to get

$$F_1(x) = \frac{F_0(0)}{(\mu_1+\lambda_2)(1-\rho)}\left[\lambda_1+\rho\mu_2-\rho(\mu_2+\lambda_1)\mathrm{e}^{-\alpha x}\right] \tag{38}$$

and

$$F_2(x) = \frac{F_0(0)}{(\mu_2+\lambda_1)(1-\rho)}\left[\lambda_2+\rho\mu_1-\rho(\mu_1+\lambda_2)\mathrm{e}^{-\alpha x}\right]. \tag{39}$$

To evaluate $F_0(0)$ we may either use the condition $\Sigma F_i(\infty)=1$ or apply renewal theory to write

$$F_0(0) = \lim_{t\to\infty} F_0(0, t) = (1+\lambda ET)^{-1}$$

where ET is given by (20).

Having obtained explicitly the joint limiting d.f. of $\{Z(t), J(t)\}$ one may obtain the marginal distributions in a straightforward manner. In particular one may check that the limiting state probabilities of $\{J(t)\}$, obtained by letting $x\to\infty$ in (35), (36), (38) and (39), agree with the results of a direct ergodic analysis of this Markov chain.

5. Some Open Problems

An objective of the present communication was to introduce to probabilists a class of outstanding problems in the area of data communication. We believe that the methods of applied probability could yield here results of significant practical and theoretical importance. In a forthcoming paper some of the results derived here will be extended in a more general model with n input lines leading into the buffer. Here we conclude with a list of some outstanding problems in this area.

(a) *Delays.* In our models we have not tackled any problems of delay. That is, how long does it take for a message to reach the main line from the moment it arrives at the buffer. When studying such problems one would, of course, have to assume a rule of priorities according to which messages are released from the buffer.

(b) *Rates of Transmission.* The assumption made here about the rates of transmission on the input lines and on the main line being equal is not always realistic. There is therefore merit in studying models in which the rate of transmission on the main line is higher than that of the input lines. Here the process $Z(t)$ may no longer be flat over any interval of time and thus it may be harder to obtain explicit expressions for the limiting distributions.

(c) *Finite Buffers.* In the models discussed here we assumed that the storage capacity of the buffer is unlimited. More sophisticated models would assume that this capacity is finite and investigate the more difficult problem of overflow.

(d) *Networks of Multiplexers.* It was already mentioned that multiplexers are usually components in data communication networks. Therefore models for multiplexers in series, in parallel or for interconnected networks of multiplexers would be of great importance.

References

1. GAVER, D. P. and LEWIS, P. A. W. Probability models for buffer storage allocation problems. *Techn. Report No.* 32, Dep. of Stat., Carnegie-Mellon University, 1969.
2. PRABHU, N. U. *Queues and Inventories.* J. Wiley, New York, 1965.
3. SMITH, W. L. Regenerative Stochastic Processes. *Proc. Roy. Soc.* (London) A, vol. **232**: 6–31, 1955.

PART III

INTERTEMPORAL PROGRAMMING: DYNAMICS

Nous ne nous tenons jamais au temps présent... : si imprudents, que nous errons dans les temps qui ne sont pas nôtres, et ne pensons point au seul qui nous appartient.... C'est que le présent, d'ordinaire, nous blesse. Nous le cachons à notre vue, parce qu'il nous afflige.... Que chacun examine ses pensées, il les trouvera toutes occupées au passé ou à l'avenir.... Le présent n'est jamais notre fin : le passé et le présent sont nos moyens; le seul avenir est notre fin. Ainsi nous ne vivons jamais, mais nous espérons de vivre; et, nous disposant toujours à être heureux, il est inévitable que nous ne le soyons jamais.

Blaise PASCAL. Pensées

CHAPTER 14

PRODUCTION PLANNING OVER TIME: SOME GENERALIZATIONS

JEAN-FRANÇOIS RICHARD

0. Introduction

This paper presents some generalizations of the algorithm derived by Hohn and Modigliani [3] to solve the following T-period problem: a firm facing known requirements wants to satisfy them at minimal cost; for each period, the planner has to decide on the amounts the firm should produce, respectively carry over. Hohn and Modigliani assumed that all costs (production and holding) were constant over time and that the marginal cost of production in each period was increasing. We will successively relax some of those assumptions.

In section 1 we will allow for varying costs over time; the Hohn-Modigliani algorithm will then appear as a special case of ours.

In section 2 we will pay more attention to an important implicit assumption in the Hohn-Modigliani solution; indeed we will see that, when there is a nonzero initial inventory, their solution could be nonoptimal in some cases, except if the disposal cost is assumed arbitrarily high. We will generalize our previous results to the case where there could be any nonnegative disposal cost.

In section 3 we will then relax the assumption of fixed requirements; the firm will be assumed to be a monopolistic one, fixing a price at each period and then selling the quantity demanded at that price. The demand function we will introduce may vary over time. Our goal will be to find a policy maximizing profits.

We will also mention some future extensions we are presently working on, considering, for example, that the firm has the option to choose (at different costs) between different plant sizes or designs.

The mathematical tool we will use is the convex program. As already pointed out by Eppen and Gould [1] in a related problem, it is a very powerful one, allowing us to derive the properties of the optimal solution and then to find a resolution algorithm. The discrete maximum principle (Halkin [2]) does not apply here because it does not allow positivity constraints on the state variables. The results we will use are found in Van Moeseke (ch. 1, section 3).

1. Varying Costs and Fixed Requirements

We define the following variables:

I_t : inventory at the end of period t $\quad t:1\rightarrow T \quad I_t \geq 0$
(in this section there is no initial inventory, an assumption to be relaxed in the next section);

P_t : quantity produced in period t $\quad t:1\rightarrow T$;

D_t : fixed requirement in period t $\quad t:1\rightarrow T$;

$F_t(P_t)$: cost of producing P_t in period t; for mathematical convenience we will assume that $F_t(P_t)$ is defined on the whole real line (but we will impose $P_t \geq 0$ on our solution); $F_t(.)$ is twice differentiable, strictly convex;

$f_t(P_t) = (\mathrm{d}/\mathrm{d}t)\, F_t(P_t)$ is the marginal cost function; $f_t(P_t)$ is then differentiable and strictly increasing;

h_t : cost per unit of holding inventory during period t; we assume that production and sales occur at a constant rate during period t; h_t is then charged on the average inventory in period t;

C_0 : fixed cost (may be the cost of building the necessary plant).

The total cost function is

$$\phi(P_t, I_t; t:1\rightarrow T) = C_0 + \sum_{t=1}^{T} F_t(P_t) + \sum_{t=1}^{T} h_t\left(\frac{I_t + I_{t-1}}{2}\right). \tag{1.1}$$

Our purpose is then to minimize ϕ subject to the constraints

$$\begin{aligned} &P_t, I_t \geq 0 \\ &P_t = D_t + I_t - I_{t-1} \quad t:1\rightarrow T \qquad (I_0 = 0). \end{aligned} \tag{1.2}$$

Having $I_0 = 0$, it is evident that the optimal solution will be such that $I_T = 0$. (Otherwise we could reduce total production and consequently costs). Let

(1.3) $$\underline{P}' = (P_1, P_2, \ldots, P_T),\ \underline{I}' = (I_1, I_2, \ldots, I_{T-1}).$$

We have

(1.4) $$\phi(\underline{P}, \underline{I}) = C_0 + \sum_{i=1}^{T} F_t(P_t) + \sum_{i=1}^{T-1} I_t\left(\frac{h_t + h_{t+1}}{2}\right).$$

ϕ is convex; it is straightforward to check that

$$\lambda\phi(\underline{P}_1, \underline{I}_1) + (1-\lambda)\phi(\underline{P}_2, \underline{I}_2) \geq \phi[\lambda(\underline{P}_1, \underline{I}_1) + (1-\lambda)(\underline{P}_2, \underline{I}_2)].$$

The strict inequality holds as soon as one of the productions involved is strictly positive.

We will assume that at least one D_t is nonzero; let us assume it is D_T (we would not affect the problem by deleting all periods following the last one characterized by a nonzero requirement: we assume no backlogs). Then there exists at least one solution with $(\underline{P}, \underline{I}) > 0$; for example if $D_t = 0$, $t: 1 \to T-1$, we could take $P_T = D_T/T$, $I_T = t(D_T/T)$. The interiority condition is then satisfied.

Defining $\underline{\Lambda} = (\lambda_1, \lambda_2 \ldots \lambda_T)$ we have

(1.5) $$L(\underline{P}, \underline{S}, \underline{\Lambda}) = C_0 + \sum_{t=1}^{T} F_t(P_t) + \sum_{t=1}^{T-1} I_t\left(\frac{h_t + h_{t+1}}{2}\right) + \sum_{t=1}^{T} \lambda_t(P_t - D_t - I_t + I_{t-1}).$$

The Kuhn-Tucker conditions are:

(1.6) $$P_t \frac{\delta L}{\delta P_t} = 0 \quad \frac{\delta L}{\delta P_t} \geq 0 \quad I_t \frac{\delta L}{\delta I_t} = 0 \quad \frac{\delta L}{\delta I_t} \geq 0 \quad \frac{\delta L}{\delta \lambda_t} = 0$$

$$t: 1 \to T \qquad t: 1 \to T-1 \qquad t: 1 \to T$$

(1.7) $$\frac{\delta L}{\delta P_t} = f_t(P_t) + \lambda_t$$

(1.8) $$\frac{\delta L}{\delta I_t} = \frac{h_t + h_{t+1}}{2} - \lambda_t + \lambda_{t+1}$$

(1.9) $$\frac{\delta L}{\delta \lambda_t} = P_t - D_t - I_t + I_{t-1}.$$

We now derive the main implications of conditions (1.6)-(1.9):

a) If the optimal solution is such that $(P_t, P_{t+k}, I_t, I_{t+1}, \ldots, I_{t+k-1}) > 0$ we must have:

$$f_t(P_t) + \lambda_t = 0$$

$$f_{t+k}(P_{t+k}) + \lambda_{t+k} = 0$$

$$\sum_{i=t}^{t+k-1} \left[\frac{h_i + h_{i+1}}{2} - \lambda_i + \lambda_{i+1}\right] = \sum_{i=t}^{t+k-1} \left(\frac{h_i + h_{i+1}}{2}\right) - \lambda_t + \lambda_{t+k} = 0$$

or:

$$\text{(I)} \qquad f_{t+k}(P_{t+k}) = f_t(P_t) + \sum_{i=t}^{t+k-1} \left(\frac{h_i + h_{i+1}}{2}\right).$$

b) If the optimal solution is such that $(P_t, P_{t+1}) > 0$, $I_t = 0$ we must have:

$$f_t(P_t) + \lambda_t = 0$$

$$f_{t+1}(P_{t+1}) + \lambda_{t+1} = 0$$

$$\frac{h_t + h_{t+1}}{2} - \lambda_t + \lambda_{t+1} \geq 0$$

or:

$$\text{(II)} \qquad f_{t+1}(P_{t+1}) \leq f_t(P_t) + \left(\frac{h_t + h_{t+1}}{2}\right).$$

c) If the optimal solution is such that $(P_t, P_{t+k}, I_t, I_{t+1}, \ldots, I_{t+k-1}) > 0$ and $P_{t+k'} = 0$, $0 < k' < k$, we find by the same procedure that we must have:

$$f_{t+k}(P_{t+k}) = f_t(P_t) + \sum_{i=t}^{t+k-1} \left(\frac{h_i + h_{i+1}}{2}\right)$$

$$\text{(III)} \qquad f_{t+k'}(P_{t+k'}) \geq f_t(P_t) + \sum_{i=t}^{t+k'-1} \left(\frac{h_i + h_{i+1}}{2}\right)$$

$$f_{t+k}(P_{t+k}) \leq f_{t+k'}(P_{t+k'}) + \sum_{i=t+k'}^{t+k-1} \left(\frac{h_i + h_{i+1}}{2}\right).$$

The two strict inequalities hold simultaneously.

Economic Interpretation

Conditions (I), (II), (III) have a straightforward economic interpretation as soon as we realize that $f_t(P_t) + \sum_{i=t}^{t+k-1} (h_i + h_{i+1})/2$ represents the marginal cost of producing one more unit in period t and carrying it into period $t+k$. It is then clear that if we had no restrictions on productions and inventories, we should respect condition (I), otherwise it would pay to reallocate the production between period t and $t+k$. But having the constraints of nonnegativity, we could have two kinds of violations:

$$(1.10) \qquad (1) \quad f_t(P_t) + \sum_{i=t}^{t+k-1} \left(\frac{h_i + h_{i+1}}{2}\right) < f_{t+k}(P_{t+k}).$$

But this clearly implies $P_{t+k} = 0$; otherwise we should produce somewhat less in period $t+k$, producing the difference in period t and carrying it into period $t+k$.

Remark. Hohn and Modigliani have no such problem because they assume that production costs are constant over time; then

$$f(P_{t+k}) > f(P_t) + \sum_{i=t}^{t+k-1} \left(\frac{h_i + h_{i+1}}{2}\right) \text{ implies } P_{t+k} > P_t \geq 0$$

and it is always possible to reduce P_{t+k}, increasing P_t by the necessary amount.

$$(1.11) \qquad (2) \quad f_t(P_t) + \sum_{i=t}^{t+k-1} \left(\frac{h_i + h_{i+1}}{2}\right) > f_{t+k}(P_{t+k}).$$

But this implies either that P_t is zero or that there exists some $0 \leq k' < k$ such that $I_{t+k'} = 0$. Otherwise we could somewhat reduce what we produce in period t and carry in period $t+k$, to increase P_{t+k} by the same amount. (This is impossible only if we do not produce in period t for period $t+k$: either $P_t = 0$ or $I_{t+k'} = 0$.)

Basic Extrapolation

DEFINITION. The basic extrapolation of (P_1, I_1) is the set $\{(P_k^*, I_k^*); k: 1 \to k_1\}$ where the (P_k^*, I_k^*) are defined by the following iterative procedure: initialization $(P_1^*, I_1^*) = (P_1, I_1)$, $P_1' = P_1$.

Step i

(1.12a) $$f_i(P'_i) = f_{i-1}(P'_{i-1}) + \tfrac{1}{2}(h_i + h_{i-1})$$

(1.12b) $$P^*_i = \max\,[0,\, P'_i]$$

(1.12c) $$I^*_i = I^*_{i-1} + P^*_i - D_i\,.$$

k_1 is defined as being the first index such that we find $I^*_{k_1} \geq 0$. If such a k_1 does not exist, we end the extrapolation in period T and we say that the basic extrapolation of (P_1, I_1) has no violations. This definition has interesting implications; we will use the notation $\{(\hat{P}_i, \hat{I}_i);\ i: 1 \to T\}$ to denote the optimal solution to our T-period problem.

If $(P_1, I_1) \geq (\hat{P}_1, \hat{I}_1)$, given the fact that conditions (I) and (II) imply that for any nonzero $\hat{P}_k$ we have

(1.13) $$f_k(\hat{P}_k) \leq f_1(P_1) + \sum_{i=1}^{k-1} \left(\frac{h_i + h_{i+1}}{2}\right)$$

and, given (1.12), we have

(1.14a) $$\hat{P}_k \leq P'_k \leq P^*_k\,.$$

The only case where we could have the reverse inequality in (1.13) would be a zero $\hat{P}_k$, but in this case we still have

(1.14b) $$0 = \hat{P}_k \leq P^*_k\,.$$

Then:

1. (P_1, I_1) can be optimal if and only if there is no violation in the basic extrapolation of (P_1, I_1); indeed if we had a k_1 such that $I^*_{k_1} < 0$, then, a fortiori given (1.14), we should have $\hat{I}_{k_1} \leq I^*_{k_1} < 0$, which is impossible in an optimal solution.

2. If the basic extrapolation of (P_1, I_1) has no violations then $\hat{P}_1 \leq P^*_1$ and consequently $\hat{P}_k \leq P^*_k$, $\forall k: 1, 2 \ldots T$. Indeed if $\hat{P}_1 > P^*_1$, we would have $\hat{I}_1 > I^*_1 \geq 0$ the (requirements are fixed); then:

$$f_2(\hat{P}_2) \geq f_1(\hat{P}_1) + \left(\frac{h_1 + h_2}{2}\right) \qquad (>: \text{only if } \hat{P}_2 = 0)$$

$$> f_1(P^*_1) + \left(\frac{h_1 + h_2}{2}\right) \geq f_2(P^*_2) \qquad \text{except for } P^*_2 = 0\,.$$

Then $\hat{P}_2 > P_2^*$ with the possible exception $\hat{P}_2 \geq P_2^* = 0$. Anyway we would have $\hat{I}_2 > I_2^* \geq 0$ and applying iteratively the same reasoning we would have $\hat{I}_k > I_k^* \geq 0$, $\forall k: 1, 2 \ldots T$, which would imply $\hat{I}_T > 0$, contradicting the fact that $\hat{I}_T$ is necessarily zero. We have a similar property: if the basic extrapolation of (P_1, I_1) has no violations and if all the implied I_k^* are strictly positive then $\hat{P}_1 < P_1^*$. Otherwise we would have $\hat{I}_k \geq I_k^* > 0$, $\forall k: 1, 2 \ldots T$.

3. If the basic extrapolation of (P_1, I_1) has a violation in k_1 $(I_{k_1}^* < 0)$, then $\hat{P}_1 > P_1^*$. Indeed as soon as $\hat{P}_1 \leq P_1^*$ we would have $\hat{P}_k \leq P_k^*$ for all k from 1 to T, implying $\hat{I}_{k_1} < 0$, which is impossible.

4. If the basic extrapolation of (P_1, I_1) has no violations and if there exists at least one k_2 $(1 \leq k_2 \leq T)$ such that $I_{k_2}^* = 0$, then $(P_k^*, I_k^*) = (\hat{P}_k, \hat{I}_k)$, $k: 1 \to k_2$. (If there is more than one k_2, we choose the greatest one.) Indeed, by condition 2) we have $\hat{P}_1 \leq P_1^*$ and then $\hat{P}_k \leq P_k^*$ $(k: 1 \to k_2)$ and as soon as one of those inequalities is strict we would have $\hat{I}_{k_2} < I_{k_2}^* = 0$, which is impossible. And if $k_2 < T$, we have $P_{k_2+1}^* > \hat{P}_{k_2+1}$. (Otherwise $\hat{I}_T$ would be strictly positive.)

Those four properties imply directly the optimality of the algorithm we now define.

Algorithm

In the first period we must at least produce $P_1 = D_1$. We then compute the basic extrapolation of $(P_1 = D_1, I_1 = 0)$.

a) If we find no violations, we have two possible cases:

   a.1) All implied I_k^* are strictly positive (and consequently $P_2^* > D_2$). As we cannot decrease P_1, $(P_1, 0)$ is optimal and we know that $\hat{P}_2 < P_2^*$. We can then consider periods 2 to T as a $(T-1)$-period problem, applying our algorithm to this new problem.

   a.2) There is at least one k_2 such that $I_{k_2}^* = 0$ (if necessary we choose the highest one); if $k_2 = T$ the basic extrapolation gives us the optimal solution; if $k_2 < T$, we have: $\{(P_k^*, I_k^*);\ k: 1 \to k_2\}$ is optimal (property 4)), and also $\hat{P}_{k_2+1} < P_{k_2+1}^*$; we can then consider the remaining $T - k_2$ periods as a new $(T-k_2)$-period problem.

b) If there is violation in period k_1, we increase continuously P_1, until we find that the basic extrapolation of $(P_1, P_1 - D_1)$ provides us with a zero $I_{k_1}^*$. (It is easy to see that if we express $I_{k_1}^*$ as a function of P_1, we have a continuous, not necessarily differentiable function of P_1.) Then we continue the basic extrapolation of the implied (P_1, I_1) beyond k_1, having two possible cases:

b.1) There are no more violations; in this case we have: $\{(P_k^*, I_k^*); k: 1 \to k_3)\}$ is optimal, k_3 being the highest index such that $I_{k_3}^* = 0$ (which exists, because we have at least $I_{k_1}^* = 0$). If $k_3 = T$ our problem is solved; if $k_3 < T$ we know that $\hat{P}_{k_3+1} < P_{k_3+1}^*$ and we consider the remaining $T - k_3$ periods as a $(T-k_3)$-period problem.

b.2) If there is a new violation in $k_1' > k_1$, we adopt for k_1' the same procedure as for k_1.

Given the properties we derived from the definition of the basic extrapolation it is clear that this algorithm provides the optimal solution. (Including a zero final inventory.)

We can also remark that the solution we find includes a certain number of zero inventories (at least $I_T^* = 0$); for the corresponding periods we have property (II) and for the other ones we have properties (I) and (III). It is clear that we can neither increase nor decrease any P_t^*:

— If we decrease one of the P_t^*, we must adapt the following productions (properties (I) and (III)). It implies that there will be at least one violation. (Some zero inventories will become strictly negative.)
— If we increase one of the P_t^* for a similar reason *all* inventories following period t become strictly positive, including the final one, and our solution is no longer optimal.

2. Problems with Nonzero Initial Inventory

The Hohn-Modigliani solution (constant costs) implicitly assumes an arbitrarily great cost of disposal. But in some situations it could be preferable to dispose of some of the initial inventory. Let us for example consider the following two-period problem:

$$D_1 = 2,\ D_2 = 6,\ F_1(x) = F_2(x) = x + 0.1\,x^2,\ h_1 = h_2 = 1,\ I_0 = 8.$$

Hohn and Modigliani assume implicitly that it is desirable to use the whole initial inventory (the only restriction being of course that this initial inventory is less than, or equal to, the total requirement). Their solution to our problem would then be $P_1^* = P_2^* = 0$, $I_1^* = 6$, $I_2^* = 0$; the total cost being

$\frac{1}{2}(8+6)+\frac{1}{2}(6+0)=10$. But if there is free disposal (or if the cost of disposal is less than 0.25 per unit) the optimal solution would be: dispose 2.5 of the initial inventory and

$$I_0 = 5.5, \quad P_1^* = 0, \quad I_1^* = 3.5, \quad P_2^* = 2.5, \quad I_2^* = 0;$$

the total cost being $\frac{1}{2}(5.5+3.5)+\frac{1}{2}(3.5)+2.5+0.1(2.5)^2 = 9.375$.

This example may appear somewhat unrealistic, but in a problem characterized by a long planning horizon (T) and a relatively high cost of holding inventories, there could be situations where the initial inventory is so important that there would be a strong incentive to dispose of some of it instead of carrying it into periods where it would be less costly to produce. This explains why we want to incorporate in our model an explicit cost of disposal, assumed to be K per unit (our model can also take into consideration a decreasing marginal cost of disposal, as we will see). The easiest way to incorporate it in a T-period problem is to transform this problem into a $(T+1)$-period problem, adding a dummy period at the beginning. This period 0 would be characterized by:

$$D_0 = 0$$

$$F_0(x) = \begin{cases} KI_0 - Kx, & x \leq I_0, \quad f_0(x) = -K, \quad x < I_0 \\ \infty \quad , & x > I_0 \end{cases} \tag{2.1}$$

$$h_0 = 0;$$

x being the quantity we do not dispose of (we will say: we "produce"); we can consider KI_0 as a fixed cost (cost we would incur if we do not want to "produce" in period 0); then "producing" x economizes the cost Kx. (If the cost of disposal is positive, it implies a negative "cost of production"; if this cost is negative we then have a positive "cost of production" equating the revenue we give up by deciding to carry rather than to dispose.)

For mathematical convenience, we will in fact use a marginal cost curve of the type illustrated by fig. 1, because we need a differentiable increasing one. ΔS, ΔC, α and β may be any arbitrary values such that the two following conditions hold:

1. The increase in the marginal cost when "production" in period 0 goes from 0 to I_0 must be negligible compared to any other relevant cost.
2. As soon as we "produce" I_0, we can bring the marginal cost to any relevant level OB by an increase in production negligible compared to the production levels we will encounter.

We can apply our algorithm as such: it implies that if, in period 0, we "produce" a quantity P, $0<P<I_0$, the inventory "carried" from period 0 into period 1 being nonzero, we must have the relation

$$f_1(P_1') = f_0(P) + \frac{h_1}{2} = -K + \frac{h_1}{2}. \tag{2.2}$$

As soon as we "produce" the whole initial inventory I_0, if there is still a violation in the basic extrapolation of period 0, we must continue to increase "production" in period 0; we know that this increase will be negligible, but it will be sufficient to raise $f_0(P)$ to a level such that, given (2.2), $f_1(P_1')$ will be sufficient to bring the necessary increases in production in periods 1, 2, ... to suppress the violation.

Practically it means that as soon as we "produce" the whole initial inventory I_0, we may disregard the "marginal cost of production in period 0" and directly raise $f_1(P_1')$ to the necessary level (equivalent to increasing P_1^* as much as necessary). Let us remark that figure 1 shows clearly why we could use a decreasing marginal cost of disposal ($f_0(P)$ must be increasing).

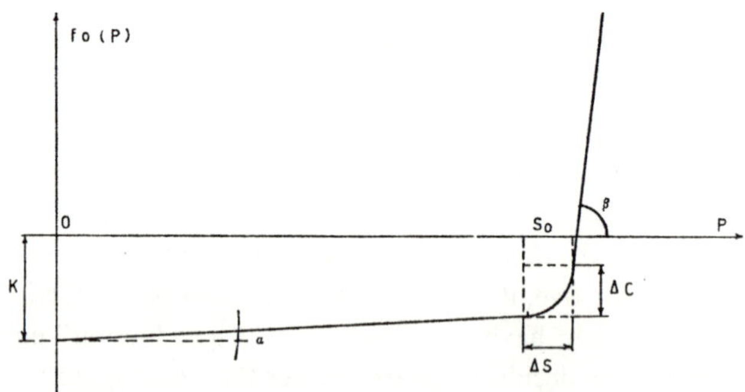

Fig. 1

Numerical Illustration

We will consider an example with a quadratic cost function; the marginal cost function is linear:

$$f_i(P) = \alpha_i + \beta_i P, \; i: 1 \to T. \tag{2.3}$$

The key relation of our basic extrapolation, $f_i(P'_i)=f_{i-1}(P'_{i-1})+\frac{1}{2}(h_i+h_{i-1})$, becomes in this case:

$$P'_i=\frac{\alpha_{i-1}-\alpha_i}{\beta_i}+\frac{\beta_{i-1}}{\beta_i}P'_{i-1}+\frac{1}{2\beta_i}(h_i+h_{i-1}). \tag{2.4}$$

The numerical application we consider is:

t	1	2	3	4	5	6
D_t	2	8	20	3	2	10
h_t	1	1	2	1	1	2
α_t	1	2	2	3	10	1
β_t	1	1	2	1	1	1

$I_0=5 \quad K=0$

(2.3) gives us the relation we will use in the basic extrapolation:

$$P'_1\begin{cases} = -0.5 & \text{if } 0\leq P'_0<5 \\ \geq \ 0.5 & \text{if } P'_0=5 \end{cases} \tag{2.5a}$$

$$P'_2=P'_1 \tag{2.5b}$$

$$P'_3=0.5\,P'_2+0.75 \tag{2.5c}$$

$$P'_4=2P'_3+0.5 \tag{2.5d}$$

$$P'_5=P'_4-6 \tag{2.5e}$$

$$P'_6=P'_5+10.5. \tag{2.5f}$$

1. We try $P'_0=0$. Then the basic extrapolation gives us:

$$P'_1=-0.5,\quad P^*_1=0,\quad I^*_1=0+0-2=-2.$$

To suppress this violation we must increase P'_0, but increasing P'_0 by less than 5 will not affect P'_1. D_1 being less than 5 will then be satisfied by P'_0:

$$P'_0=2,\quad P^*_1=0,\quad I^*_1=2+0-2=0.$$

2. The basic extrapolation of (P^*_1, I^*_1) gives:

$$P'_2=-0.5\,(=P'_1),\quad P^*_2=0,\quad I^*_2=0+0-8=-8.$$

We must still increase P'_0; $P'_0 = 5$ is not sufficient, then we continue to "increase" (negligibly) P'_0 until $f_1(P'_1)$ reaches a level such that the implied I_2^* is 0. Having $P'_2 = P'_1$, it is clear that we must choose $P'_1 = 2.5$:

$$P_0^* = 5, \qquad I_0^* = 5$$

$$P_1^* = P'_1 = 2.5, \quad I_1^* = 5+2.5-2 = 5.5$$

$$P_2^* = P'_2 = 2.5, \quad I_2^* = 5.5+2.5-8 = 0.0 .$$

3. The basic extrapolation of (P_2^*, I_2^*) gives:

$$P_3^* = P'_3 = 2.0, \quad I_3^* = 0+2-20 = -18: \text{violation.}$$

We must still increase P'_1, and given (2.5b), (2.5c) it is clear that we must increase P'_1 by $18/(1+1+0.5) = 7.2$. Then

$$P_0^* = 5, \quad I_0^* = 5, \quad P_1^* = P'_1 = 9.7, \quad I_1^* = 5+9.7-2 = 12.7$$

$$P_2^* = P'_2 = 9.7, \quad I_2^* = 12.7+9.7-8 = 14.4$$

$$P_3^* = P'_3 = 5.6, \quad I_3^* = 14.4+5.6-20 = 0 .$$

4. The basic extrapolation of (P_3^*, I_3^*) gives:

$$P_4^* = P'_4 = 11.7, \quad I_4^* = 8.7, \quad P_5^* = P'_5 = 5.7, \quad I_5^* = 12.4$$

$$P_6^* = P'_6 = 16.2, \quad I_6^* = 18.6 .$$

We know then that $\{(P_i^*, I_i^*);\ i: 1 \to 3\}$ is optimal and the last three periods may be considered as a new three-period problem.

5. We try $P_4^* = P'_4 = 3$ and then $I_4^* = 0$; the basic extrapolation of (P_4^*, I_4^*) gives:

$$P'_5 = -3, \quad P_5^* = 0, \quad I_5^* = -2 .$$

We must increase P'_4. This increase will affect P'_5, but will not affect P_5^* as long as it is less than 3. Then to set $I_5^* = 0$ we need only to increase P'_4 by 2:

$$P_4^* = P'_4 = 5, \quad I_4^* = 2$$

$$P'_5 = -1, \quad P_5^* = 0, \quad I_5^* = 0 .$$

6. The basic extrapolation of (P_5^*, I_5^*) gives:

$$P_6^* = P'_6 = 9.5, \quad I_6^* = -0.5 .$$

We still need to increase P'_4 and given (2.5*d*), (2.5*e*), (2.5*f*) we can see that we need

$$P_4^* = P_4' = 5.25, \qquad I_4^* = 2.25$$

$$P_5' = -0.75, \quad P_5^* = 0, \quad I_5^* = 0.25$$

$$P_6^* = P_6' = 9.75, \qquad I_6^* = 0.$$

The optimal solution is:

t	1	2	3	4	5	6	
P_t^*	9.7	9.7	5.6	5.25	0	9.75	
I_t^*	12.7	14.4	0	2.25	0.25	0	and $I_0^* = 5$ (no disposal).

In fact, the properties we discovered have a straightforward graphical interpretation directly related to the technique proposed by Johnson [4]: he analyzes the periods sequentially, planning the production as cheaply as possible, which is exactly what we will do graphically. It will appear that we could derive from the graphical solution an algorithm at least as simple as the one we derived above, but we keep in mind that we want to introduce prices in our model and we will see that in this case the graphical solution is inappropiate even though our algorithm will be generalized most easily.

Graphical Resolution

We assume temporarily a two-period problem, with linear marginal cost function (this last assumption is made only for graphical convenience, but any increasing function could be treated by the same procedure):

$$I_0 = 0 \begin{cases} f_1(P_1) = a_1 + b_1 P_1, & D_1, \quad h_1 \\ f_2(P_2) = \alpha_2 + \beta_2 P_2, & D_2, \quad h_2 . \end{cases} \tag{2.6}$$

To produce D_1 we have no choice: it must be in period one; but D_2 may be produced either in period two or in period one (and carried over); evidently we can also combine those two possibilities.

If we produce in period one some part x' of the requirement D_2, given that we already produced D_1 in period one and that x' must be carried over,

x' will be produced at marginal cost:

(2.7) $$a_1+b_1(D_1+x')+\tfrac{1}{2}(h_1+h_2)=\alpha_1+\beta_1 x'.$$

It means that in period two we have the choice between two marginal cost curves:

$$\begin{cases}\alpha_1+\beta_1 x \\ \alpha_2+\beta_2 x\end{cases} \quad \text{(fig. 2).}$$

(We have represented $\alpha_2<\alpha_1$, but we could in some cases have $\alpha_2 \geq \alpha_1$; the whole argument presented below would remain valid, interchanging the roles of $\alpha_1+\beta_1 x$ and $\alpha_2+\beta_2 x$.)

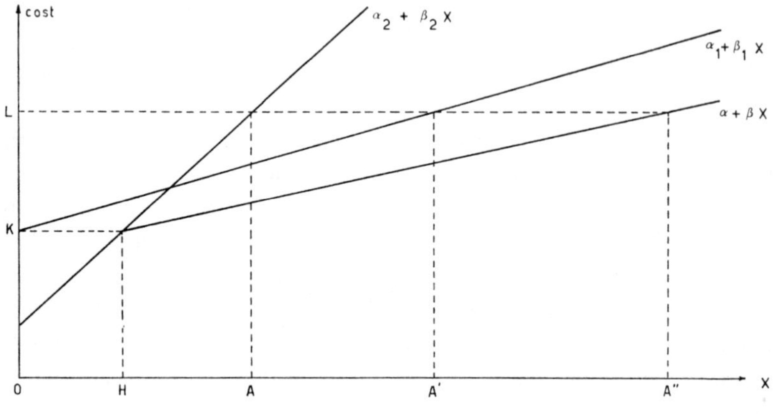

Fig. 2

If D_2 is less than OH, we will clearly produce the whole D_2 in period 2, but as soon as D_2 is greater than OH the cheapest way to produce D_2 is to divide the production between periods 1 and 2, equalizing marginal costs (which is precisely the key property of our algorithm). It means for example that, at marginal cost OL we can produce OA in period 2 and OA' in period 1 (carrying it) which gives a total production $OA''=OA+OA'$. Then the "aggregated marginal cost curve" giving us the cheapest combination of productions available in period 2 is given by:

$$\begin{cases}\alpha_2+\beta_2 x & \text{if} \quad x \leq OH \\ \alpha\ +\beta x & \text{if} \quad x > OH,\end{cases}$$

$\alpha+\beta x$ being the horizontal summation of $\alpha_1+\beta_1 x$ and $\alpha_2+\beta_2 x$.

We give here some formulas which may be useful for the graphical resolution in the linear case. Using $\alpha_2+\beta_2.OA=OL=\alpha_1+\beta_1.OA'=\alpha+\beta.OA''$, $OA=OA'+OA''$, we find

$$\alpha=\frac{\alpha_1\beta_2+\alpha_2\beta_1}{\beta_1+\beta_2} \qquad \beta=\frac{\beta_1\beta_2}{\beta_1+\beta_2}. \tag{2.8}$$

It is then easy to derive:

$$\begin{aligned} OA' &= \frac{\alpha_2-\alpha_1}{\beta_1+\beta_2}+\frac{\beta_2}{\beta_1+\beta_2}OA''=\frac{\beta_2}{\beta_1+\beta_2}HA'' \\ OA &= \frac{\alpha_1-\alpha_2}{\beta_1+\beta_2}+\frac{\beta_1}{\beta_1+\beta_2}OA''=OH+\frac{\beta_1}{\beta_1+\beta_2}HA'', \end{aligned} \tag{2.9}$$

meaning that the cheapest way to produce OA'' ($>OH$) is:

1. produce OH in period 2,
2. divide the production of the remainder HA'' between periods 1 and 2 in the proportions $\beta_1/(\beta_1+\beta_2)$ in period 1 and $\beta_2/(\beta_1+\beta_2)$ in period 2.

The extension to any number of periods is straightforward and we can treat the problem of an initial inventory exactly as we did above. To illustrate this we give now the graphical resolution of our numerical example.

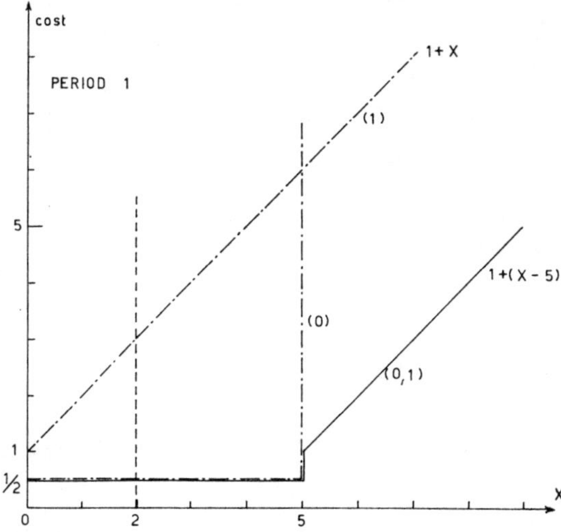

Fig. 3.1

cost
PERIOD 2
2 + X
(2)
2 + (X - 3)
(1)
2 + (X - 3)/2
(1, 2)
(0,1)
4.5
1.5
5
0
5
8
X

Fig. 3.2

cost
PERIOD 3
2 + 2X
(3)
6 + X/2
5.2 + 2/5 X
(1, 2)
((1,2), 3)
13.2
15
10
6
5
2
0
2
5
10
15
20
X

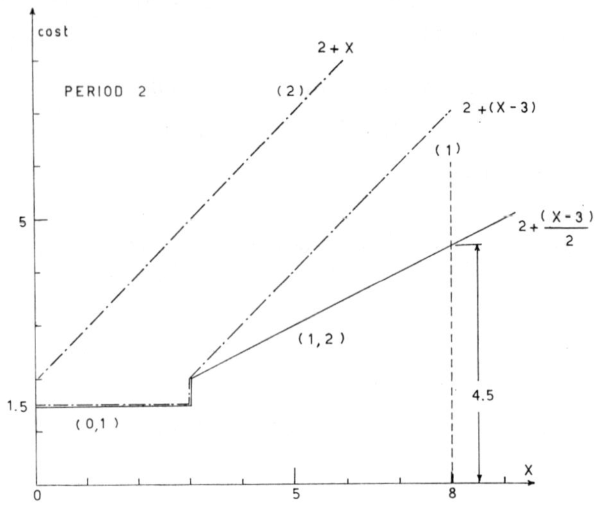

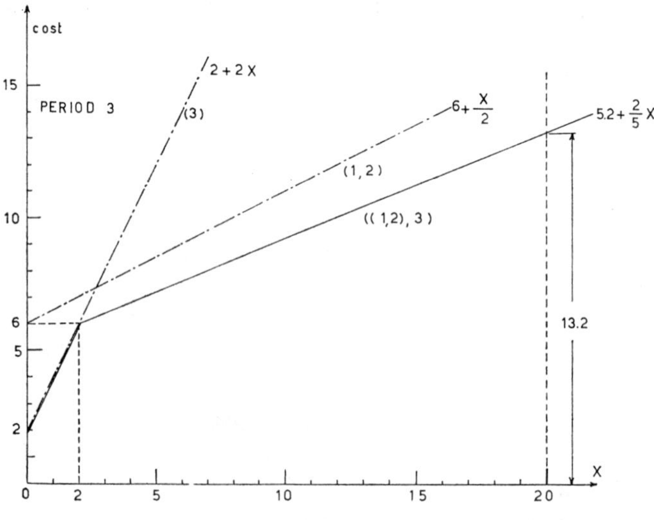

Fig. 3.3

(figs. 3.1–3.6). In those figures we adopt the following notations and conventions:

·–·–·–·–·	denotes the relevant marginal cost functions (including carrying costs)
————	denotes the "aggregate marginal cost function"
(α)	below a marginal cost curve indicates that it represents production in period α
(α, β)	indicates simultaneous production in α and β.

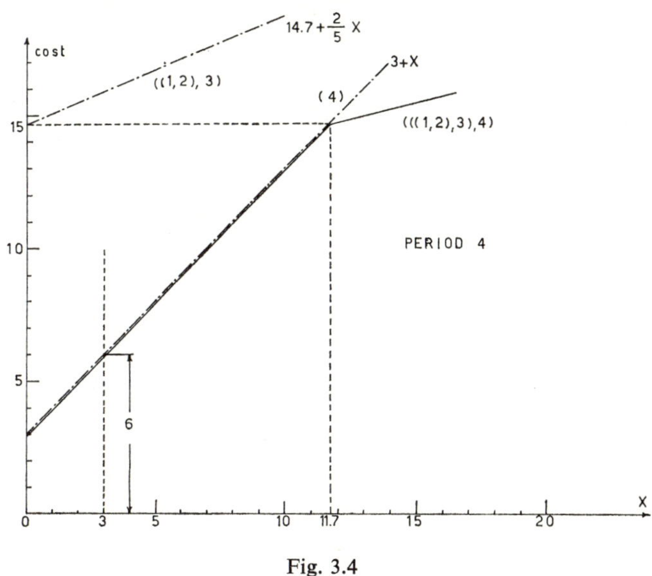

Fig. 3.4

We now briefly comment on those figures, keeping track of results in table 1.

Fig. 3.1: the requirement for period 1 is "produced" in period 0 and carried over.

Fig. 3.2: the requirement for period 2 is produced in three different periods:

3 units are "produced" in period 0 (the initial inventory is completely produced: there will be no disposal);
5 units are produced on curve (1,2) representing:

$\frac{1}{2} \cdot 5 = 2.5$ produced in period 1
$\frac{1}{2} \cdot 5 = 2.5$ produced in period 2.

Fig. 3.3: 2 units are produced on curve (3) and 18 on curve ((1, 2), 3); those 18 units are distributed as follows:

$\frac{1}{5}$ in period 3; that brings production in period 3 to $2+\frac{1}{5}\cdot 18 = 5.6$

$\frac{4}{5}$ on curve (1, 2) $\begin{cases} \frac{2}{5} \text{ in period 1, } \frac{2}{5}\cdot 18 = 7.2 \\ \frac{2}{5} \text{ in period 2, } \frac{2}{5}\cdot 18 = 7.2. \end{cases}$

Fig. 3.4: the requirement for period 4 is produced in period 4.

Fig. 3.5: the requirement for period 5 is produced in period 4.

Fig. 3.6: 9.5 units are produced on curve (6) and 0.5 on curve (4,6); this half-unit is distributed as follows:

$\frac{1}{2}$ in period 4, $\frac{1}{2}\cdot\frac{1}{2} = 0.25$

$\frac{1}{2}$ in period 6; that brings production in period 6 to $9.5+0.25 = 9.75$.

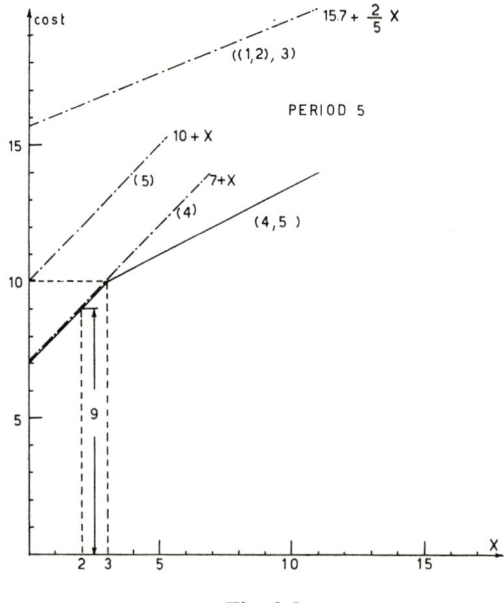

Fig. 3.5

Some Further Comments

Looking at the evolution of the solution as we proceed in our iterative search of the optimal one, we see how some periods may completely affect the tentative solution at some stage of the algorithm; but it is clear from the

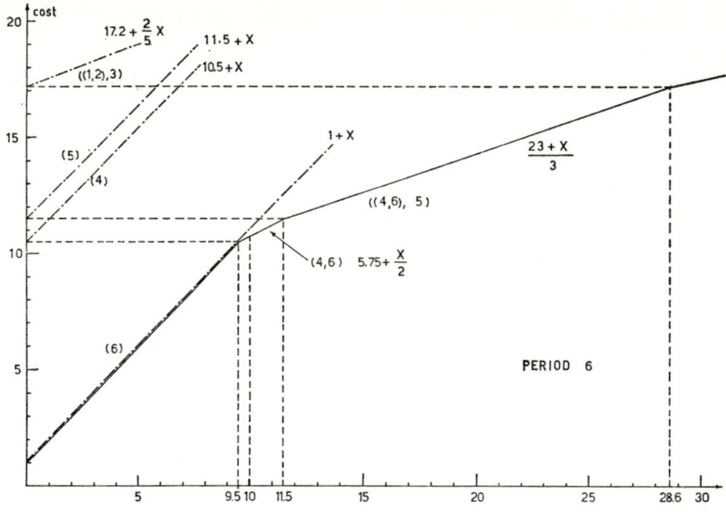

Fig. 3.6

preceding presentation that those periods are those characterized either by a sufficient increase in the requirement or by an increase in the costs of production.

On the other side, the solution implied for some subset of successive periods may also be the optimal one we would have found by considering this subset separately. In our numerical illustration periods 1, 2 and 3 were a good example of this situation; they constitute what we call a planning

TABLE 1

	PRODUCED IN PERIOD						
FOR ↓	0	1	2	3	4	5	6
1	2	0	×	×	×	×	×
2	3	2.5	2.5	×	×	×	×
3	0	7.2	7.2	5.6	×	×	×
4	0	0	0	0	3	×	×
5	0	0	0	0	2	0	×
6	0	0	0	0	0.25	0	9.75
TOTAL	5	9.7	9.7	5.6	5.25	0	9.75

horizon, characterized by the fact that the basic extrapolation of (P_3^*, I_3^*) has no violation.

Similarly, to be sure that the solution we found in a T-period problem would remain optimal in any longer problem, we do not need complete information on the periods following period T; we need only know that future costs and demands are such that there would be no violations in the basic extrapolation of (P_T^*, I_T^*); it may for example be expressed in terms of upper bounds for the possible increases.

Especially if we are planning production to meet a cyclical (growing) pattern of requirements, the preceding analysis could provide some strong justification for peak-to-peak planning; more generally it is often easy to derive some maximal rate of growth for costs and requirements such that planning peak-to-peak would in fact be optimal. We are presently investigating some interesting applications of this general problem.

3. Introduction of Prices in the Planning Algorithm

We now assume that the firm is monopolistic and that the requirements it has to satisfy depend on prices charged. More explicitly we assume demand functions instead of fixed requirements:

$$p_t = d_t(q_t), \quad t:1\to T. \tag{3.1}$$

For mathematical convenience those curves are defined on the whole real line, but q_t is constrained to the interval $0 \le q_t \le q_{Mt}$. (q_{Mt} could be infinity if we consider for example a double-log demand function.)

For each period we then have a total revenue curve:

$$R_t(q_t) = q_t \cdot d_t(q_t), \; t:1\to T. \tag{3.2}$$

We will assume that $R_t(q_t)$ is strictly concave for any $t:1\to T$, which is equivalent to assuming that the marginal revenue, $q_t d_t'(q_t) + d_t(q_t)$, is strictly decreasing:

$$2d_t'(q_t) + q_t d_t''(q_t) < 0, \quad t:1\to T. \tag{3.3}$$

Finally we assume that the firm wants to maximize total profit from period 1 to period T:

$$\max_{q_t, P_t, I_t} \sum_{t=1}^{T} [q_t d_t(q_t) - F_t(P_t)] - C_0 - \sum_{t=1}^{T-1} I_t\left(\frac{h_t + h_{t+1}}{2}\right), \tag{3.4}$$

under the constraints

(3.5 a) $$P_t + I_{t-1} = q_t + I_t \qquad t: 1 \to T$$

(3.5 b) $$P_t,\ I_t,\ q_t \geq 0 \qquad t: 1 \to T$$

(3.5 c) $$q_t \leq q_{Mt} \qquad t: 1 \to T.$$

The Lagrangean function is

(3.6) $$\sum_{t=1}^{T} [q_t d_t(q_t) - F_t(P_t)] - C_0 - \sum_{t=1}^{T-1} I_t \left(\frac{h_t + h_{t+1}}{2}\right) - \sum_{t=1}^{T} u_t(q_t - q_{Mt}) - \sum_{t=1}^{T} v_t(P_t + I_{t-1} - I_t - q_t).$$

According to the results of chapter 1, section 3, the optimality conditions are:

$$P_t \frac{\delta L}{\delta P_t} = 0, \quad \frac{\delta L}{\delta P_t} \leq 0, \quad I_t \frac{\delta L}{\delta I_t} = 0, \quad \frac{\delta L}{\delta I_t} \leq 0$$

(3.7) $$q_t \frac{\delta L}{\delta q_t} = 0, \quad \frac{\delta L}{\delta q_t} \leq 0$$

(3.8) $$u_t \frac{\delta L}{\delta u_t} = 0, \quad \frac{\delta L}{\delta u_t} \geq 0, \quad u_t \geq 0, \quad \frac{\delta L}{\delta v_t} = 0, \quad t: 1 \to T.$$

And from (3.6) we have:

(3.9 a) $$\frac{\delta L}{\delta P_t} = -f_t(P_t) - v_t$$

(3.9 b) $$\frac{\delta L}{\delta I_t} = -\left(\frac{h_t + h_{t+1}}{2}\right) + v_t - v_{t+1}$$

(3.9 c) $$\frac{\delta L}{\delta q_t} = [q_t d_t'(q_t) + d_t(q_t)] - u_t + v_t.$$

(3.9 a) and (3.9 b) imply that on the production side we have the same optimality conditions we already met in section 1; it is what we had to expect: for any set of prices, including the optimal one, if we violated one of those conditions, we could reschedule production to satisfy exactly the same requirements, prices being kept unchanged, but at a lower cost; the solution could not be optimal.

Considering the pricing policy we have essentially the following conditions; denoting by P_t^*, I_t^* and q_t^* the optimal values, if they are such that:

a) $P_t^* > 0$ and $q_{Mt} > q_t^* > 0$, then we must have $u_t = 0$ and (3.9a) (3.9c) imply

$$\text{(IV)} \qquad q_t d_t'(q_t) + d_t(q_t) = f_t(P_t).$$

b) $P_t^* = 0$ and $q_{Mt} > q_t^* > 0$, then q_t^* must be produced in some preceding periods; be $t-k$ ($k>0$) one of those periods; this implies that I_{t-k}^*, $I_{t-k+1}^*, \ldots, I_{t-1}^*$ are strictly positive; consequently we have:

$$\text{(3.9a')} \qquad f_{t-k}(P_{t-k}) = -v_{t-k}$$

$$\text{(3.9b')} \qquad \sum_{i=t-k}^{t-1} \left(\frac{h_i + h_{i+1}}{2} \right) = v_{t-k} - v_t$$

$$\text{(3.9c')} \qquad [q_t d_t'(q_t) + d_t(q_t)] = -v_t$$

and

$$\text{(V)} \qquad f_{t-k}(P_{t-k}) + \sum_{i=t-k}^{t-1} \left(\frac{h_i + h_{i+1}}{2} \right) = [q_t d_t'(q_t) + d_t(q_t)].$$

This type of equality for any period $t-k$ ($P_{t-k}^* > 0$) relates to period t by nonzero inventories (and given condition I, there are no ambiguities).

c) $q_t^* = 0$, then conditions IV and V become respectively:

$$\text{(IV')} \qquad q_t d_t'(q_t) + d_t(q_t)|_{q_t=0} \leq f_t(P_t)$$

$$\text{(V')} \qquad \leq f_{t-k}(P_{t-k}) + \sum_{i=t-k}^{t-1} \left(\frac{h_i + h_{i+1}}{2} \right).$$

d) $q_t^* = q_{Mt}$, then $u_t \geq 0$ and condition (3.9c′) becomes $q_t d_t'(q_t) + d_t(q_t) \geq -v_t$, condition which can be related to the different possible cases for P_t^*, for example if $P_t^* \neq 0$: $q_t d_t'(q_t) + d_t(q_t)|_{q_t = q_{Mt}} \geq f_t(P_t)$.

If $I_t^* > 0$, $q_{t+1}^* > 0$, $P_t^* > 0$:

$$q_t d_t'(q_t) + d_t(q_t)|_{q_t = q_{Mt}} + \left(\frac{h_t + h_{t+1}}{2} \right) \geq f_t(P_t) + \left(\frac{h_t + h_{t+1}}{2} \right)$$

$$\text{(VI)} \qquad = q_{t+1} d_{t+1}'(q_{t+1}) + d_{t+1}(q_{t+1})|_{q_{t+1} = q_{t+1}^*}.$$

Economic Interpretation

All those conditions have a straightforward economic interpretation; roughly stated: the quantities sold are priced accordingly to the principle:

marginal revenue = marginal cost, but the relevant marginal cost is the effective cost of producing the quantities sold in period t, which can include carrying costs if the production comes from a period preceding the t-th one.

We may have two possible exceptions (inequalities):

a) A zero sale in period t is the result of some combination of the two following situations:
   - due to some future increase in demand, the marginal revenue the firm can obtain by carrying the production of period t is higher than the marginal revenue it can obtain now, even at the maximum marginal revenue (first unit which would be sold in period t);
   - some future increase in production costs pushes production in period t to a level such that selling in period t would be less profitable.

b) The cost of production in period t (or in the period where we produce for period t) is so low that the firm would be willing to sell more than q_{Mt} in period t (but consumers do not want to buy more than q_{Mt}); in this case it is clear that we may have inequality (VI).

We can now extend the algorithm defined in section 1.

Basic Extrapolation

DEFINITION. The extrapolation of (P_1, I_1) is the set $\{(P_k^*, I_k^*); k : 1 \to k_1\}$ where the (P_k^*, I_k^*) are defined by the following iterative procedure: initialization $(P_1^*, I_1^*) = (P_1, I_1)$.

Step i. 1) Compute P_k^* by

$$f_i(P_i') = f_{i-1}(P_{i-1}') + \tfrac{1}{2}(h_i + h_{i-1}) \tag{3.10}$$

$$P_i^* = \max\,[0, P_i']. \tag{3.11}$$

2) Compute q_k^* and consequently I_k^* by

$$[d_k(q_k') + q_k' \cdot d_k'(q_k')] = f_k(P_k'), \tag{3.12}$$

which means: if $P_k' > 0$, $P_k^* = P_k'$ and we use (IV)

if $P_k' \leq 0$, $P_k^* = 0$ and we use (V).

$$q_k^* = \max\,[0, q_k']; \tag{3.13}$$

eventually also q_k^* must be constrained by q_{Mt}:

$$I_k^* = I_{k-1}^* + P_t^* - q_k^*. \tag{3.14}$$

Algorithm with Zero Initial Inventory

1. We check for optimality of $I_1 = 0$: we successively compute

$$q_1' \text{ by } [q_1' d_1'(q_1') + d_1(q_1')] = f_1(q_1')$$
$$P_1^* = q_1^* = \max[0, q_1'] \tag{3.15}$$
$$I_1^* = 0.$$

2. We then derive the basic extrapolation of $(P_1^*, 0)$.

2a) If there is no violation, we look at the greatest k_2 such that $I_{k_2}^* = 0$ (it always exists because we have $I_1^* = 0$) and $\{(P_t^*, I_t^*); t: 1 \to k_2\}$ is optimal; we then consider the remaining periods as a new $(T-k_2)$-period problem.

2b) If there is a violation in k_1 ($I_{k_1}^* < 0$) we must increase P_1^*; for any P_1^* we try, we can compute the optimal associated I_1^* as follows:

$$q_1' \text{ is computed by } [q_1' d_1'(q_1') + d_1(q_1')] = f_1(P_1^*)$$
$$q_1^* = \max[0, q_1'] \tag{3.16}$$
$$I_1^* = P_1^* - q_1^*.$$

(We see that the effect of an increase in P_1^* is accentuated by the consequent decrease in q_1^*.) We increase P_1^* until we find that the basic extrapolation of (P_1^*, I_1^*) provides us a zero $I_{k_1}^*$. Then we continue the basic extrapolation of (P_1^*, I_1^*) beyond k_1; if there is no more violation, we choose the highest k_3 such that $I_{k_3}^* = 0$ and $\{(P_t^*, I_t^*); t: 1 \to k_3\}$ is optimal; we consider the remaining periods as a new $(T-k_3)$-period problem. If there is a new violation in k_3 ($k_3 > k_1$) we apply again procedure 2b), replacing k_1 by k_3.

In fact, the introduction of pricing in our model magnifies the effects of variations in production. For example, an increase in production in period k will require higher prices in some following periods and consequently lower requirements in those periods (given the optimality conditions we derived). The implied effects on inventories are enforced and the scheme of the demonstration we gave in section 1 remains completely valid. The reader can easily incorporate the slight necessary adaptations in this demonstration.

Extension to a Nonzero Initial Requirement

As in section 2 we introduce a dummy period 0, the production function having the characteristics defined there. We assume any demand function having the required properties, but characterized by a marginal revenue curve so low relatively to the other ones that it will never be optimal to "sell" in period zero. (We prefer this formulation to a similar one assuming any demand function having the required properties but imposing $q_{M0}=0$; it is essentially for mathematical convenience, having to satisfy the interiority and Kuhn-Tucker conditions.)

Numerical Application

We will consider a linear situation

$$d_t(q_t) = \lambda_t - v_t q_t, \quad t:1 \to T \tag{3.17}$$

$$f_t(P_t) = \alpha_t + \beta_t P_t, \quad t:1 \to T. \tag{3.18}$$

To insure a positive price we need $\lambda_t/v_t \geq q_t$ and define $q_{Mt} = \lambda_t/v_t$; but this constraint is no problem: as soon as we produce in a period, the optimality condition will clearly imply that we sell at some positive price; we will not consider situations where we have such an important initial inventory and a high disposal cost (are both really compatible!) that we would be ready to subsidize consumers to avoid disposal. (In any case in such a situation the linearity of the demand function would be most unrealistic.)

The marginal revenue is given by

$$\frac{\delta}{\delta_{q_t}} [q_t(\lambda_t - v_t q_t)] = \lambda_t - 2v_t q_t. \tag{3.19}$$

For convenience we repeat here the relation we will use in the basic extrapolation:

$$P'_{t+1} = \frac{\alpha_t - \alpha_{t+1}}{\beta_{t+1}} + \frac{\beta_t}{\beta_{t+1}} P'_t + \frac{1}{2\beta_{t+1}} (h_t + h_{t+1}). \tag{3.20}$$

The numerical application we consider is

t	1	2	3	4	5	6	
λ_t	10	4.9	11.44	3	3.9	8	
v_t	1	0.75	0.2	1	0.5	0.75	
α_t	3	2	5	2	10	2	
β_t	1	0.5	1	0.5	1	0.4	
h_t	0.5	0.3	0.5	0.4	0.2	0.4	$I_0 = 7,\ K=0.$

1. We try $P'_0 = 0$ ($\to I_0 = 0$) which corresponds to $f_0(P'_0) = 0$. The basic extrapolation of (P'_0, I_0) gives us:

   $f_1(P'_1) = 0.25, \quad P'_1 = -2.75, \qquad P^*_1 = 0$

   $10 - 2q'_1 = 0.25, \; q'_1 = 4.875 = q^*_1, \; I^*_1 = -4.875$, violation in period 1.

   We must increase P'_0 enough to have the implied $I^*_1 = 0$; increasing P'_0 does not affect P'_1 as long as $P'_0 < 7$. We try:

   $P'_0 = 4.875 = P^*_0, \qquad I^*_0 = 4.875$

   $P'_1 = -2.75, \; P^*_1 = 0, \; q^*_1 = 4.875, \; I^*_1 = 0 \; (f_1(P'_1) = 0.25)$;

   so if we had one period problem we would dispose 2.125 of the initial inventory and

(3.21) $$P^*_1 = 0, \quad q^*_1 = 4.875.$$

2. The basic extrapolation of (P^*_1, I^*_1) gives:

   $$f_2(P'_2) = 0.65, \qquad P'_2 = -2.7, \quad P^*_2 = 0$$

   $$4.9 - 1.5q'_1 = 0.65, \quad q^*_2 \cong 2.83, \quad I^*_2 \cong -2.83.$$

   We must still increase P'_0, for the same reason that in 1 it is clear that here we would need $P'_0 = 7$ $(4.875 + 2.83 > 7)$; we now know that we will not dispose of our initial inventory. Increasing P'_0 by a negligible amount we can raise $f_0(P'_0)$ to any necessary level. Increasing $f_1(P'_1)$ will have a double effect:

   — $f_1(P'_1)$ and $f_2(P'_2)$ will increase and this may be sufficient to induce some production

   — but prices will increase and the requirements will consequently decrease.

   Instead of trying different successive values for $f_1(P'_1)$, we may try a parametric search. Let $f_1(P'_1) = \theta$ (we know $\theta > 0.25$); this implies successively:

   $$f'_2(P'_2) = \theta + 0.4$$

   (given f_i, if $\theta < 3.00$ no production in 1
   $\theta < 1.60$ no production in 2)

   $$10 - 2q'_1 = \theta, \qquad q'_1 = 5 - \frac{\theta}{2}$$

   $$4.9 - 1.5q'_2 = \theta + 0.40, \quad q'_2 = 3 - 2\frac{\theta}{3}.$$

First try $\theta<1.60$ (no production); the clearance condition gives:

$$I_2^* = 7 - \left(5 - \frac{\theta}{2}\right) - \left(3 - 2\frac{\theta}{3}\right) = 0, \quad \theta = \frac{6}{7} < 1.6\,.$$

(If we find $\theta>1.6$, we must adapt the clearance condition to the fact that for $3>\theta>1.6$ we produce in 2 and for $\theta>3$ we produce in 1 and 2.) If we had a two-period problem the solution would be

$$\text{(3.22)} \qquad \text{no disposal} \begin{cases} P_1^* = 0, \quad q_1^* = q_1' = \dfrac{32}{7}, \quad I_1^* = \dfrac{17}{7} \\[2ex] P_2^* = 0, \quad q_2^* = q_2' = \dfrac{17}{7}, \quad I_2^* = 0 \end{cases}$$

$$\left(f_2(P_2') = \frac{6}{7} + 0.4 = \frac{44}{35}\right).$$

3. The basic extrapolation of (P_1^*, I_1^*) gives:

$$f_3(P_3') = \frac{44}{35} + \frac{2}{5} = \frac{58}{35}, \quad P_3'<0, \quad P_3^* = 0$$

$$11.44 - 0.4q_3' = \frac{58}{35}, \qquad q_3'>0, \quad I_3^* < 0, \text{ violation.}$$

A parametric analysis similar to the preceding one shows successively:

1) we must choose $\theta>1.6$,
2) we must choose $\theta>3.0$;

then adapting the clearance condition we find $\theta=5$. If we had a three-period problem the solution would be:

$$\text{(3.23)} \qquad \text{no disposal} \begin{cases} f_1(P_1') = 5, \quad P_1^* = P_1' = 2, \quad q_1^* = q_1' = 2.5, \quad I_1^* = 6.5 \\ f_2(P_2') = 5.4, \quad P_2^* = P_2' = 6.8, \quad q_2' < 0, \quad q_2^* = 0, \quad I_2^* = 13.3 \\ f_3(P_3') = 5.8, \quad P_3^* = P_3' = 0.8, \quad q_3^* = q_3' = 14.1, \quad I_3^* = 0. \end{cases}$$

It is trivial to check that if we continue the basic extrapolation we find no more violation: periods 1, 2 and 3 constitute a planning horizon and we have a brand new problem starting in period 4.

4. Let us try whether $I_4^* = 0$ may be optimal; we then must have marginal revenue equal to marginal cost:

$$(3.24)\quad \begin{aligned} &2+0.5q_4' = 3-2q_4' \rightarrow P_4' = q_4' = P_4^* = q_4^* = 0.4 \\ &f_4(P_4') = 3-0.8 = 2.2\,. \end{aligned}$$

5. But the basic extrapolation of (P_4^*, I_4^*) has a violation $I_5^* = -1.4$. We must increase P_4'; we try parametrically:

$$P_4' = \theta,\ f_4(P_4') = 2+0.5\theta,\ q_4' = 0.5-0.25\theta \quad (q_4'>0 \text{ if } \theta<2)$$

if $\theta<2$: $I_4^* = 1.25\theta-0.5,\quad f_5(P_5') = 2.3+0.5\theta$

$$P_5' = -7.7+0.5\theta,\quad P_5^* = 0,\quad q_5' = 1.6-0.5\theta>0$$

and $I_5^* = 0 \rightarrow 1.25\theta-0.5 = 1.6-0.5\theta\,.$

$\theta = 1.2$ which is less than 2. If we had found a θ greater than 2 we would have had $q_4^* = 0$ and some modification of the clearance condition.

If we had a five-period problem, the solution for the last two periods would have been

$$(3.25)\quad \begin{aligned} &P_4^* = 1.2, \quad q_4^* = q_4' = 0.2, \quad I_4^* = 1 \\ &P_5^* = 0, \quad q_5^* = q_5' = 1, \quad I_5^* = 0, \quad f_5(P_5') = 2.9\,. \end{aligned}$$

6. The basic extrapolation of (P_4^*, I_4^*) has a violation $I_6^*<0$; a similar parametric analysis allows us to find the optimal solution for the last three periods:

$$(3.26)\quad \begin{aligned} &P_4^* = 1.26, \quad q_4^* = 0.185, \quad I_4^* = 1.075, \quad f_4(P_4') = 2.63 \\ &P_5^* = 0, \quad q_5^* = 0.970, \quad I_5^* = 0.105, \quad f_5(P_5') = 2.93 \\ &P_6^* = 3.075, \quad q_6^* = 3.180, \quad I_6^* = 0, \quad f_6(P_6') = 3.23\,. \end{aligned}$$

We solved this application in detail because it has two interesting features: the first one is purely formal, illustrating the kinds of problems we may face in a practical analysis. The second one is economic: it is most interesting to follow the evolution of the solution as we add successive periods ((3.21) to (3.26)) and to see how the monopolist "exploits" the situation: in fact it will carry inventories from periods characterized by a low cost of production into periods where costs are higher: it really penalizes consumers belonging to such a period of low cost (period 2 is a striking example).

Somebody could object there are few situations where a monopolist is completely autonomous for his pricing policy. But the methodology we used is very flexible and can easily take into account additional restrictions.

Graphic Presentation

We now briefly explain the problem one would encounter when trying to generalize the graphical resolution we gave in section 2 (and, more generally, to extend Johnson's solution).

We assume a two-period linear problem

$$\begin{array}{lll} \text{period 1:} & \text{marginal revenue} & \lambda_1 - 2v_1 q \\ & \text{marginal cost} & \alpha_1 + \beta_1 q \\ \text{period 2:} & \text{marginal revenue} & \lambda_2 - 2v_2 q \\ & \text{marginal cost} & \alpha_2 + \beta_2 q . \end{array}$$

If we were planning period by period, in period 1 we would produce and sell $P_1^* = q_1^* = (\lambda_1 - \alpha_1)/(\beta_1 + 2v_1) = OQ_1$ (fig. 4.1) and the marginal cost would be $OR_1 = (\beta_1 \lambda_1 + 2v_1 \alpha_1)/(\beta_1 + 2v_1)$.

Adding period 2, our preceding graphical resolution would suggest that we have the choice between two marginal cost curves:

$\alpha_2 + \beta_2 q$,

$OR_2 + \beta_1 q$ with $OR_2 = OR_1 + \frac{1}{2}(h_1 + h_2)$,

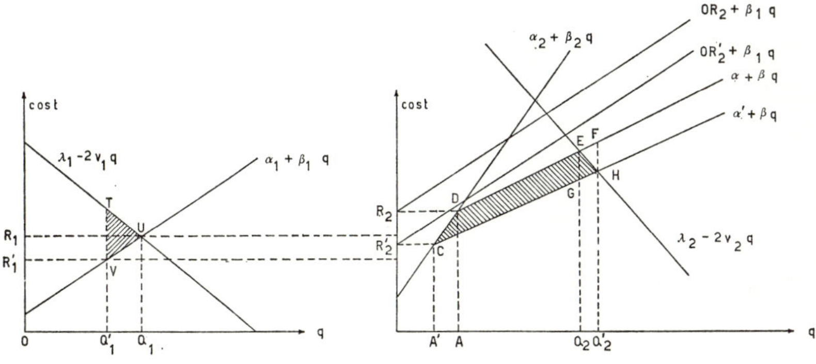

Fig. 4.1 Fig. 4.2

which provides us with an "aggregate marginal cost curve" $\alpha'+\beta q$ for $q>OA$ and we would conclude that the optimal production in periods 1 and 2 for period 2 should be $OQ_2=(\lambda_2-\alpha)/(\beta+2v_2)$. But the introduction of prices reinforces the interrelations between periods. To show this let us assume that in period 1 the monopolist sells only $OQ_1'=OQ_1-\phi$; he looses a profit equal to the shaded area TUV (fig. 4.1); but now, OR_1 becomes OR_1' and there is a shift in some curves in fig. 4.2. Having reduced his production in period 1 for period 1, the monopolist can now produce at a lower cost in period 1 for period 2:

$$OR_2 \text{ becomes } OR_2' = OR_1' + \tfrac{1}{2}(h_1+h_2)$$

$$\alpha \quad \text{becomes } \alpha' .$$

It is easy to see that in period 2 the monopolist gains a profit equal to the shaded area $CDEH$.

As soon as $CDEH>ABC$ the monopolist has an incentive to reduce OQ_1. In fact it is easy to obtain ABC and $CDEH$ in function of ϕ and to show that as soon as OA is less than OQ_2 there is an optimal ϕ strictly positive. (The condition $OA<OQ_2$ is also exactly the condition for having a positive inventory from period 1 into period 2 if we solve the problem algebraically; one has $I_1=\beta_2/(\beta_1+\beta_2)\,A'Q_2'$. But this is only a two-period problem, and in a larger one the interrelations between periods become much more intricate (a loss of profit in period 1 brings a gain in periods 2, 3, ... but a loss in period 2 brings a gain in period 3, ... which affects the profitability of the loss in period 1, ...).

4. Concluding Remarks

We are convinced that the preceding analysis has two most interesting aspects.

1. The first one is to illustrate how the Kuhn-Tucker procedure is powerful in some dynamic problems, a result already emphasized by Eppen and Gould. The main reason is that most of the dynamic problems we could encounter in the theory of the firm are such that the solution may be expressed recursively:

$$I_1 = \phi(P_1)$$

$$P_2 = Q(I_1, P_1) \ldots$$

The Kuhn-Tucker condition allows us to derive properties such that we can solve the whole problem conditionally by a small number of decisions (P_1, for example, in our algorithm); the problem then reduces to the choice of those variables.

2. The second one is to be flexible enough to allow further needed extensions. We know indeed that the static distinction between short-term and long-term policy for the firm is somewhat deficient as soon as the demand faced by the firm fluctuates. In fact, if there are no uncertainties about those fluctuations the firm has essentially two options to smooth their impact: carry inventories or adapt plant size (or design). A study of this problem is not complete as long as it does not include both aspects simultaneously; for example, the choice of plant size is strongly related to the nature of the fluctuations but also to the carrying cost.

We are presently analyzing the problem of a monopolistic firm facing a cyclical growth pattern of demand and some evolution in costs due, for example, to technological progress; we try to derive the implications of an optimal policy at two simultaneous levels:

— how to plan production at the seasonal level, and how to derive the optimal-holding inventory policy,
— given some cost of adapting plant size (or design), cost related to technological progress, for example, what is the optimal (dynamic) investment policy of the firm?

Preliminary results seem promising. Another generalization will be uncertainty and its effects on the optimal solution.

References

1. EPPEN, G. D. and GOULD, F. J. A Lagrangian application to production models. *Operations Research*, **16**: 819–829, 1968.
2. HALKIN, A. A maximum principle of the Pontryagin type for the systems described by nonlinear difference equations. *S.I.A.M. Journal on Control*, **4**: 90–111, 1966.
3. HOHN, F. E. and MODIGLIANI, F. Production planning over time and the nature of the planning horizon. *Econometrica*, **23**: 46–66, 1955.
4. JOHNSON, S. M. Sequential production planning over time at minimum cost. *Management Science*, **3**: 435–438, 1957.

CHAPTER 15

ON MODELS OF ECONOMIC GROWTH WITH RANDOM ELEMENTS

TAMAR DAR, ASSAF RAZIN, and JOSEPH A. YAHAV

0. Introduction

The stochastic dynamic programming model of Economic Growth in the one-commodity case was considered by several authors: Phelps [4], Levhari and Srinivasan [3] (L & S for short), Samuelson [5], Brock and Mirman [1]. In this paper we are elaborating on some aspects of this model that did not get much attention by previous authors.

We use the setup given by L & S [3], namely: at each period t, an individual (or planner) has the option of either consuming all his capital k_t, or investing part of it. Denoting consumption at time t by c_t, investment would be $k_t - c_t$ (we assume $0 \leq c_t \leq k_t$ throughout this paper). This investment will result in his capital in the next period k_{t+1} becoming $f_{r_t}(k_t - c_t)$ where, for any given $r_t, f_{r_t}(\cdot)$ would be called the production function. We take r_t to be random. The planner has an instantaneous utility function $U(\cdot)$, where the argument is the consumption at time t. The objective of the planner is to maximize

$$E\{\sum_{t=0}^{\infty} \beta^t U(c_t)\}, \tag{0.1}$$

where β is the subjective discount factor (we assume $0<\beta<1$) and $E(\cdot)$ the expectation operator.

We assume throughout this paper, unless otherwise stated, that $f_{r_t}(\cdot)$ and $U(\cdot)$ are twice differentiable and that

(0.2) $\quad U'(\cdot)>0, \; U''(\cdot)<0, \qquad \lim_{x \to 0} U'(x) = \infty;$

(0.3) $\quad f'_{r_t}(\cdot)>0, \; f''_{r_t}(\cdot)\leq 0, \qquad$ for each r_t a.s.

(0.4) $\quad \{r_t : t = 1, 2, \ldots\}$ are independent and identically distributed elements (for short i.i.d.).

It can be shown that, under these assumptions, a stationary solution exists and is given by a consumption function that for each k strictly satisfies the constraints, namely

$$0 < C(k) < k. \tag{0.5}$$

In section 1 we consider the limiting behavior of k_t in two special cases not covered in Brock and Mirman [1]. The cases are:

(i) (0.6) $$f_r(x) = rx, \quad U(x) = \frac{x^{1-\alpha}}{1-\alpha}, \quad \alpha > 0, \quad r \geq 0,$$

(ii) (0.7) $$U(x) = \lg x, \quad f_{r,\alpha}(x) = rx^{\alpha}, \quad 0 < \alpha \leq 1, \quad 0 \leq r.$$

In these cases the solution is $C(k) = \lambda k$ and we are able to express k_t in an explicit form which enables us to investigate directly the behavior of k_t.

In section 2 we consider the problem of random "death" and its effect on the optimal consumption function. We will develop the model in the above setup as follows:

$U(\cdot)$ satisfies (0.2) and

$$f_{r_t, T}(\cdot) = \begin{cases} f(\cdot), & t < T \\ 0, & t \geq T \end{cases} \tag{0.8}$$

where T is a random time and $f(\cdot)$ satisfies (0.3). A similar model was considered by Yaari [6]. We consider here the effect of "age" on optimal consumption.

In section 3 we consider aggregate consumption of individuals where the individuals operate in a model given by (0.6) and (0.7). We assume that different individuals differ in their initial capital k_0 and by their subjective discount factor β. We compare the optimal consumption behavior to consumption in an egalitarian society.

1. Limiting Behavior of k_t

The limiting behavior of k_t is of interest both when we consider the optimal behavior of the individual and even more so when we consider aggregation of individuals. The random "death" model makes capital k_T (T is the random death time) the amount that is left over to the new generation so when we consider T large the limiting behavior of k_t plays an important role.

In the deterministic case, under some assumption, there exists a steady state and $k_t \to k^*$ as $t \to \infty$ where k^* is the steady state. We want to investigate the problem in the stochastic case.

We will consider two special cases which are not covered by Brock and Mirman [1].

Case (*i*). We repeat (0.6) of the introduction:

(1.1) $$f_r(x) = rx, \quad 0 \le r$$

(1.2) $$U(x) = \frac{x^{1-\alpha}}{1-\alpha}, \quad 0<\alpha, \quad 1 \ne \alpha$$

(1.3) $$r_i \text{ are i.i.d. } \quad E[r^{1-\alpha}] < \frac{1}{\beta}, \quad P(r \ge 0) = 1\,.$$

L & S [3] considered the optimization problem here and got the following solution: the optimal consumption function is linear, i.e. $C(k) = \lambda k$, where

(1.4) $$\lambda = 1 - \beta^{1/\alpha} E^{1/\alpha}(r^{1-\alpha})\,.$$

This solution determines k_t as follows:

(1.5) $$k_t = (1-\lambda)^t k_0 \prod_{i=1}^{t} r_i\,.$$

In order to analyze k_t we will take logs on both sides and we get

(1.6) $$\lg k_t = \lg k_0 + \sum_{i=1}^{t} \{\lg r_i + \lg(1-\lambda)\}\,.$$

By a theorem of Chung and Fuchs [2] the behavior of $\lg k_t$ depends only on $E[\lg r + \lg(1-\lambda)]$, namely

(1.7) $$E[\lg r] + \lg(1-\lambda) < 0 \Leftrightarrow \lg k_t \to -\infty \quad \text{a.s.}$$

(1.8) $$E[\lg r] + \lg(1-\lambda) > 0 \Leftrightarrow \lg k_t \to \infty \quad \text{a.s.}$$

Using (1.4) the left hand side of (1.7) is equivalent to

(1.9) $$\alpha E[\lg r] + \lg E[r^{1-\alpha}] < -\lg \beta\,.$$

On the other hand the right hand side of (1.7) is equivalent to

(1.10) $$k_t \to 0 \quad \text{a.s.}$$

Combining (1.7), (1.9) and (1.10) we get

(1.11) $$k_t \to 0 \text{ a.s.} \Leftrightarrow \alpha E[\lg r] + \lg E[r^{1-\alpha}] < -\lg \beta;$$

(1.3) and (1.11) are consistent if

(1.12) $$E[\lg r] < 0$$

and hence

(1.13) $$E[\lg r] < 0 \Rightarrow k_t \to 0 \text{ a.s.}$$

Since $E[r] < 1 \Rightarrow E[\lg r] < 0$, hence

(1.14) $$E[r] < 1 \Rightarrow k_t \to 0 \text{ a.s.}$$

In an analogous way we get that the left hand side of (1.8) is equivalent to

(1.15) $$\alpha E[\lg r] + \lg E[r^{1-\alpha}] > -\lg \beta.$$

On the other hand the right hand side of (1.8) is equivalent to

(1.16) $$k_t \to \infty \text{ a.s.}$$

Combining (1.8), (1.15) and (1.16) we get

(1.17) $$k_t \to \infty \text{ a.s.} \Leftrightarrow \alpha E[\lg r] + \lg E[r^{1-\alpha}] > -\lg \beta.$$

(1.3) and (1.17) imply

(1.18) $$k_t \to \infty \text{ a.s.} \Rightarrow E[\lg r] \geq 0$$

and hence,

(1.19) $$k_t \to \infty \text{ a.s.} \Rightarrow E[r] \geq 1.$$

A special case is the case of equality, i.e.

(1.20) $$\alpha E[\lg r] + \lg E[r^{1-\alpha}] = -\lg \beta.$$

In this case the almost sure behavior of k_t is that of oscillation on the ray $(0, \infty)$. If we assume that

(1.21) $$V[\lg r] < \infty$$

we can use the Central Limit Theorem to conclude

(1.22) $$\frac{\lg k_t}{\sqrt{tV(\lg r)}} \xrightarrow{D} N(0, 1),$$

where $\overset{D}{\to}$ denotes convergence in distribution and $N(0, 1)$ denotes the standard normal distribution.
(1.20) and (1.3) imply

$$E[\lg r] > 0 \tag{1.23}$$

and hence (1.22) and (1.3) imply

$$E[r] > 1. \tag{1.24}$$

Case (*ii*). We repeat (0.7):

$$U(X) = \lg x, \tag{1.25}$$

$$f_{r_t \alpha_t}(x) = r_t x^{\alpha t}, \tag{1.26}$$

(1.27) $\{r_i\}$ and $\{\alpha_i\}$ are sequences of i.i.d. variables and we assume the sequence $\{r_t\}$ to be independent of the sequence $\{\alpha_t\}$;

$$P(0<\alpha\leq 1) = 1,\ P(r\geq 0) = 1. \tag{1.28}$$

We compute the optimal solution following the steps used by L & S to solve case (ii). The solution satisfies

$$U'(C(k)) = \beta E\{U'(C(f_{r,\alpha}(K-C(k)))\cdot f'_{r\alpha}(K-C(k))\}, \tag{1.29}$$

which reduces under our assumptions to

$$\frac{1}{C(k)} = \beta E \frac{\alpha(k-C(k))^{\alpha-1} r}{C(r(k-C(k))^{\alpha})}. \tag{1.30}$$

$C(k) = \lambda \cdot k$ satisfies (1.30), where

$$\lambda = 1 - \beta E[\alpha]. \tag{1.31}$$

Now we can write explicitly the expression for k_n:

$$k_n = \prod_{i=1}^{n} r_i^{\Pi_{j=i+1}^{n} \alpha_j} \cdot (1-\lambda)^{\Sigma_{i=1}^{n} (\Pi_{j=i}^{n} \alpha_j)} \cdot k_0^{\Pi_{i=1}^{n} \alpha_i} \tag{1.32}$$

and hence

$$\lg k_n = \sum_{i=1}^{n} (\prod_{j=i+1}^{n} \alpha_j) \lg r_i + \sum_{i=1}^{n} (\prod_{j=i}^{n} \alpha_j) \lg (1-\lambda) + (\prod_{i=1}^{n} \alpha_i) \lg k_0. \tag{1.33}$$

In order to determine the behavior of $\lg k_n$ we will first prove a lemma.

LEMMA 1.1. Let $\{X_i: i=1, 2\ldots\}$ be i.i.d. and let $\{Y_i: i=1, 2\ldots\}$ be i.i.d., where the sequence $\{Y_i\}$ is independent of the sequence $\{X_i\}$; define

$$Z_n = \sum_{i=1}^{n} y_i \prod_{j=i+1}^{n} X_j \qquad \left(\prod_{j=n+1}^{n} X_j \equiv 1\right). \tag{1.34}$$

If $$P(0<X_i \leq 1) = 1 \quad \text{and} \quad E[X] < 1$$

then Z_n converges in distribution.

We will prove the lemma here under the additional assumption that $E[|J|] < \infty$.

Proof. Let

$$\tilde{Z}_n = \sum_{i=1}^{n} Y_i \prod_{j=1}^{i-1} X_j. \tag{1.35}$$

Notice that $\tilde{Z}_n$ and Z_n have the same distribution.

We will show that $\tilde{Z}_n \to \tilde{Z}$ a.s. where $\tilde{Z}$ is a random variable.

Consider

$$\tilde{Z}_{n+k} - \tilde{Z}_n = \sum_{i=n+1}^{n+k} Y_i \prod_{j=1}^{i-1} X_j = \left(\prod_{j=1}^{n} X_j\right) \sum_{i=n+1}^{n+k} Y_i \prod_{j=n+1}^{i-1} X_j. \tag{1.36}$$

Let

$$W_{n,k} = \tilde{Z}_{n+k} - \tilde{Z}_n. \tag{1.37}$$

We will show that W_{nk} converges to zero a.s. uniformly in k.

$$P\{|W_{n,k}| \geq l\} \leq \frac{E|W_{nk}|}{l} \leq \frac{E|Y| \cdot (E[X])^n}{l(1-E[X])} \tag{1.38}$$

and hence

$$\tilde{Z}_n \to \tilde{Z} \text{ a.s.} \tag{1.39}$$

We have to show that

$$\operatorname*{Lim}_{z\to\infty} P(\tilde{Z} \leq z) = 1 \quad \text{and} \quad \lim_{z\to-\infty} P(\tilde{Z} \leq z) = 0 \tag{1.40}$$

$$P(\tilde{Z} \leq z) = P\left(\sum_{i=1}^{\infty} Y_i \prod_{j=1}^{i-1} X_j \leq z\right) \tag{1.41}$$

and hence for $z>0$

$$P(\tilde{Z} > z) \leq P\left(\sum_{i=1}^{\infty} |Y_i| \prod_{j=1}^{i-1} X_j > z\right) \leq \frac{E[|Y|]}{z(1-E[X])} \tag{1.42}$$

and for $z<0$

(1.43) $$P(\tilde{Z}\leq z)=P(-\tilde{Z}\geq z)\leq\frac{E[|Y|]}{-z(1-E[X])}$$

so that (1.42) and (1.43) imply (1.40). Since $\tilde{Z}_n$ and Z_n have the same distribution we can conclude that the sequence Z_n has a limiting distribution which is equal to the distribution of $\tilde{Z}$. QED.

Applying lemma 1.1 to the first and second sums on the right hand of (1.33), and noticing that the third term converges to zero, we get that $\lg k_n$ converges in distribution and hence so does k_n. Lemma 1.1 does not cover the case where r might take the value zero with a positive probability. In such a case it is easily seen from (1.32) that $k_n\to 0$ a.s. Lemma 1.1 also does not cover the case where α is equal to 1 almost surely; in this case we are back to our analysis of case (i) and we get

(1.44) $$k_n\to 0 \text{ a.s.} \Leftrightarrow E[\lg r]<-\lg\beta$$

hence

(1.45) $$E[r]<\frac{1}{\beta}\Rightarrow k_n\to 0 \text{ a.s.}$$

(1.46) $$k_n\to\infty \text{ a.s.} \Leftrightarrow E[\lg r]>-\lg\beta$$

hence

(1.47) $$k_n\to\infty \text{ a.s.} \Rightarrow E[r]>\frac{1}{\beta}.$$

If $$E[\lg r]=-\lg\beta \quad\text{and}\quad V(\lg r)<\infty$$

(1.48) $$\frac{\lg k_n}{\sqrt{nV(\lg r)}}\xrightarrow{D} N(0,1).$$

2. Random "Death" and its Effect on the Consumption Function

In this section we will develop a model of optimal consumption under uncertainty of lifetime. We will assume

(2.1) $$U(0)=0,\ U'(\cdot)>0,\ U''(\cdot)<0,\ U'(0)=\infty$$

(2.2) $$f_{r_t,T}(\cdot)=\begin{cases} f_{r_t}(\cdot), & t\leq T\\ 0\ , & t>T\end{cases}$$

where f satisfies

(2.3) $f_{r_t}(0)=0,\ f'_{r_t}(\cdot)>0,\ f''_{r_t}(\cdot)\leq 0$ for each r_t, $\{r_t: t=1,2,\ldots\}$

are i.i.d., and T, r_t are independent.

It is useful to start the analysis by considering the optimal model in which the horizon is not a random variable and the discount factor depends on time, i.e., we have a sequence $(\beta_0, \beta_1 \ldots)$ where β_t is the discount factor in period t. Later we will show that the random "death" model can be reduced to this model.

The problem is:

$$(2.4)\qquad V_\beta(k,0)=\max_{\{c_t\}} E\sum_{t=0}^{\infty}\Big(\prod_{i=0}^{t}\beta_i\Big)U(c_t),\ \beta_0=1,$$

$$0\leq\beta_t\leq 1,\quad \underline{\beta}=(\beta_0,\beta_1,\ldots)$$

with the stochastic constraints $0\leq c_t\leq k_t$, $k_{t+1}=f_{r_t}(k_t-c_t)$.

We first consider the finite-horizon case. Let $V_{\underline{\beta}}^N$ denote the maximized expected value of the sum of discounted utilities, and let $C_{\underline{\beta}}^N(k)$ denote the optimal initial consumption, when the length of the horizon is N. From dynamic programing we have

$$(2.5)\qquad V_{\underline{\beta}}^N(k_0)=U(c_{\underline{\beta}}^N(k_0))+\beta_1 EV_{\underline{\beta}}^{N-1}(f_r(k_0-c_{\underline{\beta}}^N(k_0))),$$

where $\underline{\beta}=(\beta_2,\beta_3\ldots\beta_N)$.

Following L & S [3] we can characterize the optimum solution by

$$(2.6)\quad U'(c_{\underline{\beta}}^N(k_0))=\beta_1 E\left\{\frac{\mathrm{d}}{\mathrm{d}k}V_{\underline{\beta}}^{N-1}(f_r(k_0-c_{\underline{\beta}}^N(k_0)))\,f_r'(k_0-c_{\underline{\beta}}^N(k_0))\right\}$$

$$(2.7)\qquad \frac{\mathrm{d}}{\mathrm{d}k}V_{\underline{\beta}}^N(k_0)=U'(c_{\underline{\beta}}^N(k_0)),$$

where prime denotes derivative. It is well-known that $V_{\underline{\beta}}^N(k)$ is concave and differentiable.

We will prove a lemma for the finite-horizon case.

LEMMA 2.1. Let $\underline{\beta}=(\beta_0,\beta_1,\ldots,\beta_N)$ and $\underline{\beta}^*=(\beta_0^*,\beta_1^*,\ldots\beta_N^*)$ be two sequences of discount factors, and assume $\beta_t^*\geq\beta_t$ for $t=0,1,\ldots N$ then $c_{\underline{\beta}^*}^N(k)\leq c_{\underline{\beta}}^N(k)$.

Proof (by induction on N).

For $N=0$, $c_{\underline{\beta}}^0(k)=c_{\underline{\beta}^*}^0(k)=k$.

Assume now that for $N-1$, $c_\beta^{N-1}(k) \geq c_{\beta^*}^{N-1}(k)$ for all $k \geq 0$.

To show that for N the same inequality holds, assume there exists a k such that $c_\beta^N(k) < c_{\beta^*}^N(k)$.

Consider now the two sides of equation (2.6) when β^* substitutes for β. The right hand side increases for the following reasons. For any r

$$f_r'(k - c_{\beta^*}^N(k)) \geq f'(k - c_\beta^N(k)) \tag{2.8}$$

by hypothesis. Also, $f_r(k - c_{\beta^*}^N(k)) < f_r(k - c_\beta^N(k))$ for any r implies

$$\frac{\mathrm{d}}{\mathrm{d}k} V_\beta^{N-1}(f_r(k - c_\beta^N(k))) \leq \frac{\mathrm{d}}{\mathrm{d}k} V_\beta^{N-1}(f_r(k - c_{\beta^*}^N(k))) \tag{2.9}$$

by the concavity of V_β^{N-1}. Denote

$$f_r(k - c_{\beta^*}^N(k)) = k_1 , \tag{2.10}$$

then from (2.7) we have

$$\frac{\mathrm{d}}{\mathrm{d}k} V_\beta^{N-1}(k_1) = U'(c_\beta^{N-1}(k_1)) . \tag{2.11}$$

Using the induction hypothesis for $N-1$, the concavity of U, and (2.7) we get

$$U'(c_\beta^{N-1}(k_1)) \leq U'(c_{\beta^*}^{N-1}(k_1)) = \frac{\mathrm{d}}{\mathrm{d}k} V_{\beta^*}^{N-1}(k_1) . \tag{2.12}$$

Observing (2.9)–(2.12) we have

$$\frac{\mathrm{d}}{\mathrm{d}k} V_\beta^{N-1}(f_r(k - c_\beta^N(k))) \leq \frac{\mathrm{d}}{\mathrm{d}k} V_{\beta^*}^{N-1}(f_r(k - c_{\beta^*}^N(k))) \text{ for every } r; \tag{2.13}$$

since $\beta_1^* \geq \beta_1$, (2.8) and (2.13) imply

$$\beta_1 E \frac{\mathrm{d}}{\mathrm{d}k} V_\beta^{N-1}(f_r(k - c_\beta^N(k))) f_r'(k - c_\beta^N(k)) \leq \tag{2.14}$$

$$\leq \beta_1^* E \frac{\mathrm{d}}{\mathrm{d}k} V_{\beta^*}^{N-1}(f_r(k - c_{\beta^*}^N(k))) f_r'(k - c_{\beta^*}^N(k)) .$$

In other words the right hand side of (2.6) increases (or remains unchanged) upon substituting β^* for β. On the other hand the strict concavity of u

implies that the left hand side of (2.6) decreases, thus equality (2.6) is violated.

Therefore we must have, for all k,

$c^N_{\underline{\beta}}(k) \geq c^N_{\underline{\beta}^*}(k)$. This completes the proof of the lemma. QED.

In the infinite-horizon case we denote by $c_{\underline{\beta}}(k)$ the optimal initial consumption and we assume that $V_{\underline{\beta}}(k)$ in (2.4) is bounded. The counterpart of lemma 2.1 in this case is the following lemma.

LEMMA 2.2. Let $\underline{\beta} = (\beta_0, \beta_1, \ldots)$ and $\underline{\beta}^* = (\beta^*_0, \beta^*_1, \ldots)$ be two infinite sequences of discount factors and assume $\beta^*_t \geq \beta_t$ for $t = 0, 1, 2, \ldots$, then $c_{\underline{\beta}^*}(k) \leq c_{\underline{\beta}}(k)$.

Proof. First observe that $c^N_{\underline{\beta}}(k)$ (the optimal initial consumption for the N-period problem) is monotonically decreasing in N. To see this consider the sequences $(\beta_0, \beta_1, \ldots \beta_N, 0)$ and $(\beta_0, \beta_1, \ldots \beta_N, \beta_{N+1})$. Since the second sequence is larger than or equal to the first one elementwise, it follows from lemma 2.1 that

$$c^{N+1}_{(\beta_0 \ldots \beta_N \beta_{N+1})}(k) \leq c^{N+1}_{(\beta_0 \beta_1, \ldots \beta_N 0)}(k) = c^N_{(\beta_0 \ldots \beta_N)}(k).$$

This implies that $\bar{c}_{\underline{\beta}}(k) = \lim_{N \to \infty} c^n_{\underline{\beta}}(k)$ exists.

Since $V^N_{\underline{\beta}}(k)$ is monotonically increasing in N and is bounded by $V_{\underline{\beta}}(k)$ (the maximand of the infinite-horizon problem) it has a limit $\overline{V}_{\underline{\beta}}(k)$,

$$\overline{V}_{\underline{\beta}}(k) \leq V_{\underline{\beta}}(k). \tag{2.15}$$

On the other hand let $\{c_t\}^{\infty}_{t=0}$ be the stochastic solution of the infinite horizon case. Thus $\{c_t\}^N_{t=0}$ (the truncated sequence) is a feasible stochastic solution for any N, which implies that

$$V^N_{\underline{\beta}}(k) \geq E \sum_{t=0}^{N} \prod_{i=0}^{t} \beta_i U(c_t). \tag{2.16}$$

Taking limits in both sides of (2.16) yields

$$\overline{V}_{\underline{\beta}}(k) \geq V_{\underline{\beta}}(k). \tag{2.17}$$

Therefore (2.15) and (2.17) imply

$$V_{\underline{\beta}}(k) = \overline{V}_{\underline{\beta}}(k). \tag{2.18}$$

To show that $\bar{c}_{\underline{\beta}}(k)$ is an optimal initial consumption in the infinite-horizon case, we consider equation (2.5). It is legitimate to take limits in both sides of equation (2.5) since $V_\beta^N(k)$ and $c_\beta^N(k)$ have limits, u is continuous, and $V_\beta^{N-1}(f_r(k-c_\beta^N(k)))$ converges monotonically (thus, the limit of the expected value is equal to the expected value of the limit).

Thereby we get, using (2.18),

$$V_\beta(k)=u(\bar{c}_\beta(k))+\beta_1 EV_\beta(f_r(k-\bar{c}_\beta(k))). \tag{2.19}$$

Equation (2.19) implies that $\bar{c}_\beta(k)=c_\beta(k)$ is an optimal consumption in the infinite horizon problem.

From lemma 2.1 we know that $c_{\beta*}^N(k)\leq c_\beta^N(k)$ for all N, thus $c_{\beta*}(k)\leq c_\beta(k)$. This completes the proof. QED.

We return now to the model of optimal consumption under uncertainty of lifetime given by (2.1)–(2.3).

Since the sum of the discounted utilities is now a random variable we will maximize the expectation of this variable over the distribution of T and r_t. We have

$$V^T(k)=\max_{\{c_t\}} E_{T,r}\sum_{t=0}^{\infty}\beta^t U(c_t) \text{ subject to } 0\leq c_t\leq k_t \text{ for all } t, \text{ and}$$

$$k_{t+1}=f_{(r_t,T)}(k_t-c_t).$$

Let $Z(k_0)$ denote the expected value for the discounted sum of utilities for a feasible stochastic solution $\{c_t\}$, since r_t and T are independent:

$$Z(k_0)=\sum_{N=0}^{\infty}P(T=N)E_r\sum_{t=0}^{N}\beta^t U(c_t).$$

Since $U(\cdot)\geq 0$ we can interchange the order of summation to get

$$Z(k_0)=E_{r_t}\{\sum_{t=0}^{\infty}\sum_{N=t}^{\infty}P(T=N)\beta^t U(c_t)\}=E_r\sum_{t=0}^{\infty}P(T\geq t)\beta^t U(c_t), \tag{2.20}$$

where $P(T\geq t)=\sum_{N=t}^{\infty}P(T=N)$ is the probability of surviving until t. Define now the probability of death at time t conditional on surviving $t-1$ periods by $P_t=P(T=t,\ T\geq t)$ so that

$$P(T\geq t)=(1-P_0)(1-P_1)\dots(1-P_{t-1}).$$

We can see then that the model of random "death" is equivalent to the infinite-horizon model with a sequence of discount factors $\beta_0, \beta_1, \dots$, which depends on time in the following way:

$$\beta_t=\beta(1-P_{t-1}),\ t\geq 1,\ \beta_0=1. \tag{2.21}$$

Substituting (2.21) into (2.20) we get

$$Z(k_0) = E_r \sum_{t=0}^{\infty} \left(\prod_{i=0}^{t} \beta_i\right) U(c_t).$$

In the standard infinite-horizon model ($\beta_t = \beta$), a stationary solution is obtained. In the random "death" case, the solution, in general, will not be stationary except in the special case where the conditional probability of death time does not depend on age, i.e., $P_t = P$, $t = 1, 2, \ldots$ Then, (2.21) reduces to

$$\beta_t = \beta(1-P),$$

which leads to the standard infinite-horizon model with a smaller discount factor.

We examine now the effect of "age" on optimal consumption. In order to investigate the pure effect of aging we will analyze optimal consumption as a function of age, holding the amount of capital constant.

Assume that the conditional probability of death increases with age, which means

$$P_t \uparrow \text{ with } t.$$

Then $\beta_t = \beta(1-P_t)$ decreases with t.

Denote by $c_t(k)$ the optimal consumption which corresponds to the sequence of discount factors $(\beta_t, \beta_{t+1}, \ldots)$. A straightforward application of lemma 2.2 implies that the optimal consumption functions $c_t(k)$ are increasing with t.

We therefore conclude that, when age increases the conditional probability of death (holding constant the amount of capital), it will increase consumption.

3. Aggregate Consumption

In sections 1, 2 we dealt with an individual (or planner) whereas in this section we consider a "community".

We usually consider β to be subjective for an individual; thus for the community we have a distribution function over β. We will simplify our analysis by assuming $U(\cdot)$, the utility function, to be the same over the individuals. Our purpose is to consider different community policies and compare the results.

We will assume first that we are in case i of section 1 and that (1.1), (1.2) and (1.3) hold. Suppose that the community is using the services of a planner and this community is an egalitarian one; i.e., the constraints imposed on the planner require equal consumption. Therefore the objective function is given by

$$\max E\left[\sum_{t=0}^{\infty} \beta^t U(C_t)\right], \tag{3.1}$$

where the expectation is taken over the production function and the random variable β. (In this case β is independent of r.)

The constraints are given by

$$0 \leq c_t(m) \leq m_t\,, \tag{3.2}$$

where m_t is the average capital of the community and $c_t(m)$ is the consumption of each individual at time t.

We can write

$$V(m) = U[c(m)] + \sum_i \{E[\beta_i V(f_r(m-c(m)))]\}\, P(\beta_i) \tag{3.3}$$

and hence

$$V(m) = U(c(m)) + \bar{\beta} E[V(f_r(m-c(m))], \tag{3.4}$$

where $\bar{\beta}$ is the average β in the community.

This model is reduced to the previous one discussed in section 1 and hence the solution is given by

$$c_t(m) = \bar{\lambda} m_t\,, \tag{3.5}$$

where

$$\bar{\lambda} = 1 - \bar{\beta}^{1/\alpha} E^{1/\alpha}[r^{1-\alpha}]. \tag{3.6}$$

We will compare $c_t(m)$, given in (3.5), to the average consumption if the community did not impose the egalitarian consumption constraint. In the latter case each individual will behave optimally according to his β and also his k would differ. The average consumption of the community can be defined by

$$\tilde{c}_t(\tilde{k}) = \sum_i \lambda_{\beta_i} k_i P(\beta_i, k_i), \tag{3.7}$$

where $\tilde{k}$ denotes the amount of capital of the community, $\tilde{c}$ the average consumption and $P(\beta_i, k_i)$ the weighting of individual i. Using (1.4) we get

$$\tilde{c}_t(\tilde{k}) = \sum_i \{1-\beta_i^{1/\alpha} E^{1/\alpha}[r^{1-\alpha}]\} k_i P(\beta_i, k_i). \tag{3.8}$$

Equation (3.8) is now compared with (3.5) and we have

$$\tilde{c}_t(\tilde{k}) - c_t(m) = E^{1/\alpha}[r^{1-\alpha}] \{\bar{\beta}^{1/\alpha} m - \sum_i \beta_i^{1/\alpha} k_i P(\beta_i, k_i)\}. \tag{3.9}$$

If $\alpha<1$ and we assume that the k_i are the same over individuals, i.e. $k_i = m$ we see that

$$\tilde{c}_t(\tilde{k}) - c_t(m) < 0. \tag{3.10}$$

If $\alpha>1$ and we assume that $k_i = m$ then

$$\tilde{c}_t(\tilde{k}) - c_t(m) > 0. \tag{3.11}$$

In case $\alpha<1$ (3.10) indicates that the egalitarian consumption is larger.

We note that individuals with high β consume less than individuals with small β having the same k; so, as a result, k_i and β_i may be expected to be positively correlated (assuming that the "starting point" of all individuals was the same). In this case we can conclude that

$$\sum_i \beta_i^{1/\alpha} k_i P(\beta_i, k_i) > (\sum_i \beta_i^{1/\alpha} P(\beta_i)) \sum_i k_i P(k_i) = m \sum_i \beta_i^{1/\alpha} P(\beta_i) \tag{3.12}$$

and so (3.10) holds also without assuming $k_i = m$. Unfortunately we cannot conclude that (3.11) holds universally.

Analyzing case (ii) of section 1 in the same fashion as above we get

$$\bar{\lambda} = 1 - \bar{\beta} E\alpha \tag{3.13}$$

$$\tilde{c}_t(\tilde{k}) = \sum_i (1-\beta_i E\alpha) k_i P(\beta_i, k_i) = m - \sum_i \beta_i E\alpha k_i P(\beta_i, k_i). \tag{3.14}$$

If we assume $k_i = m$ we get

$$\tilde{c}_t(\tilde{k}) - c_t(m) = 0. \tag{3.15}$$

Otherwise, with a positive correlation between β_i and k_i

$$\tilde{c}_t(\tilde{k}) - c_t(m) = E\alpha\{\bar{\beta} m - \sum_i \beta_i k_i P(\beta_i, k_i)\} < 0. \tag{3.16}$$

In this case (3.16) indicates that egalitarian consumption is larger than the aggregate consumption in a community which does not impose the equal-consumption constraint.

References

1. BROCK, W. A. and MIRMAN, L. J. Optimal economic growth and uncertainty: The discounted case. *Journal of Economic Theory* (forthcoming).
2. CHUNG, K. L. and FUCHS, W. H. J. On the distribution of values of sums of random variables. *Mem. Amer. Math. Soc.*, **6**: 1–12.
3. LEVHARI, D. and SRINIVASAN, T. N. Optimal savings under uncertainty. *Review of Economic Studies*, **35**: 153–163, 1968.
4. PHELPS, E. S. The accumulation of risky capital: A sequential utility analysis. *Econometrica*, **30**: 729–743, 1962.
5. SAMUELSON, P. A. Lifetime portfolio selection by dynamic stochastic programming. *Review of Economics and Statistics*, **31**: 239–246, 1969.
6. YAARI, M. E. Uncertain lifetime, life insurance, and the theory of the consumer. *Review of Economic Studies*, **32**: 137–150, 1965.

CHAPTER 16

OPTIMAL DESIGN OF A ROTATING DISK

ETIENNE LOUTE

Summary

The minimum-weight design of axially symmetric, rotating elastic disks subject to various constraints is formulated as an optimal-control problem. After quantifying the different variables the problem is treated via dynamic programming. Given the state and control variable at stage i, the state variables at state $i+1$ are computed using a special approximation. Numerical examples are presented to show the weak sensitivity of the method to the quantification of the state variables and to the approximation in their computation.

1. Introduction

The optimal-design problem we consider in this paper is the weight minimization of a turbine disk which is a rotating part of a jet engine. For the sake of simplicity, the turbine disk is regarded as a circular rotating disk of variable thickness. It may have a hole at the center, and support blades around the periphery. When the engine works, the temperature can be high and varies in function of the disk radius. A disk must bear stresses due to rotation (influence of centrifugal forces), stresses due to external loads (the blades) and thermal stresses. For safety reasons the stresses should be below some level and for all sorts of reasons a special shape can be required for some section of the disk. The first constraints are called safety or behavioral constraints while the second ones are called geometric constraints.

In first approximation we can consider only two main stresses in the disk: the radial stress and the circumferential or tangential stress. These stresses may be regarded as functions of the radius only. Be

μ the specific mass of the metal;

ν the Poisson ratio ($\nu = 0.3$);
$\varepsilon(r)$ the expansion coefficient versus radius;
$E(r)$ the modulus of elasticity versus radius;
ω the angular speed;
r_0 the outer radius;
r_i the inner radius;
$\sigma_r(r)$ the radial stress versus radius;
$\sigma_t(r)$ the tangential stress versus radius;
σ_{r0} the outer radial stress;
$w(r)$ the thickness versus radius;
$t(r)$ the temperature versus radius.

The stresses are solutions of a system of two differential equations with two limit conditions:

(1) $$\frac{\mathrm{d}}{\mathrm{d}r}\{w(r)\,r\sigma_r(r)\} - w(r)(\sigma_r(r) - \mu\omega^2 r^2) = 0$$

(2) $$(1+\nu)\left(\frac{\sigma_t(r) - \sigma_r(r)}{E(r)}\right) + r\frac{\mathrm{d}}{\mathrm{d}r}\left\{\frac{\sigma_t(r) - \nu\sigma_r(r)}{E(r)} + \varepsilon(r)\,t(r)\right\} = 0$$

(3) $$\sigma_r(r_i) = 0 \qquad \sigma_r(r_0) = \sigma_{r0}\,.$$

2. The Optimization Problem

Given μ, ν, ε, E, r_i, r_0, σ_{r0} and $t(r)$, the optimization problem is to find a function $w(r)$ such that the disk has minimum weight and that some behavioral and geometrical constraints (fig. 1) are satisfied [5].

Geometrical Constraints

$$w_m(r) \leq w(r) \leq w_M(r), \quad r_i \leq r \leq r_0$$

$$\frac{\mathrm{d}w}{\mathrm{d}r}(r) \text{ is given in some subintervals of } [r_i, r_0]\,.$$

Behavioral Constraints

Given: $\sigma_{rM}(r)$ upper bound of $\sigma_r(r)$ versus radius
$\sigma_{tM}(r)$ upper bound of $\sigma_t(r)$ versus radius;

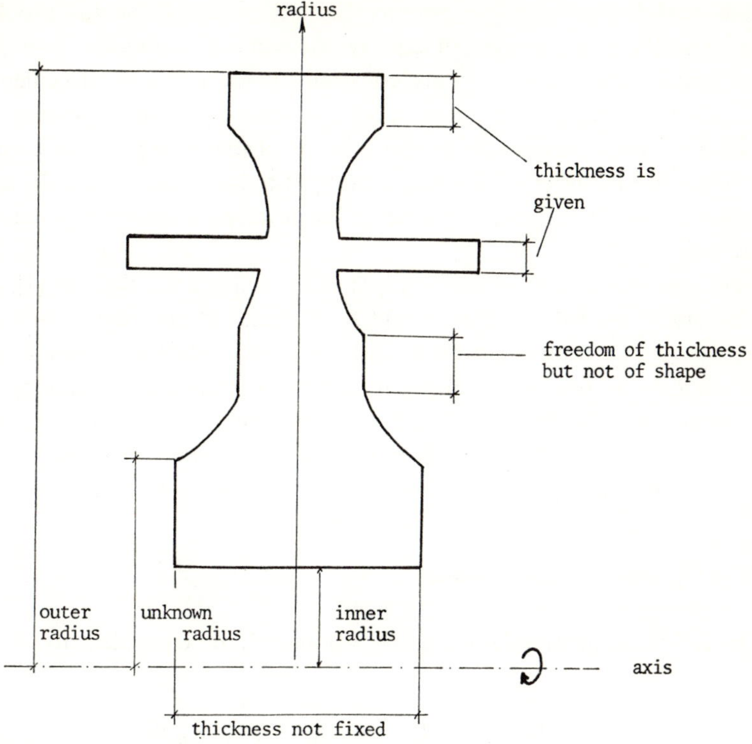

Fig. 1

the constraints are:

$$\max_{r_i \leq r \leq r_0} (\sigma_r(r) - \sigma_{rM}(r)) \leq 0$$

$$\max_{r_i \leq r \leq r_0} (\sigma_t(r) - \sigma_{tM}(r)) \leq 0.$$

A similar problem, but less constrained, has been treated by de Silva [3]. The problem is reformulated as a nonlinear programming problem with nonanalytic constraints for the behavioral constraints. A minimization procedure for solving problems with nonanalytic constraints is developed. All the constraints presented in [5] cannot be taken into account in the procedure. For instance, shape constraints cannot be treated. Since in this model the profile of the disk is defined by a finite number of points which are design variables, computer time for solving the problems increases rapidly with the number of those variables.

One can also consider the problem as a variational-calculus problem. Chern and Prager [2] derive a necessary and sufficient optimality condition for a simple problem whose only constraint is the radial displacement of edge, which is given. In a recent paper Di Stefano [4] using dynamic programming gives a remarkably simple solution of the same problem in terms of a linear partial differential equation subject to initial conditions. In order to treat a more constrained problem, he develops a two-sweep iterative procedure.

Since the problem we consider [5] is more constrained than in [4], and in a different way, we are led to consider a discrete optimal-control problem. Discrete optimal-control problems can be solved either by dynamic programming or by mathematical programming. Dynamic programming can be used to solve the disk problem, for the discrete optimal-control problem can be formulated with only three state variables and only one control variable. Problems of this order are still tractable by computer.

3. The Optimal-Control Problem

The following differential equations (4) and (5) are equivalent to (1) and (2):

$$\frac{d}{dr}\sigma_t(r) = -\nu\frac{dw}{dr}(r)\frac{\sigma_2(r)}{w(r)} - \frac{1}{r}(\sigma_t(r)-\sigma_r(r)) - \varepsilon E\frac{dt}{dr}(r) - \nu\mu\omega^2 r \tag{4}$$

$$\frac{d}{dr}\sigma_r(r) = -\frac{dw}{dr}(t)\frac{\sigma_r(r)}{w(r)} + \frac{1}{r}(\sigma_t(r)-\sigma_r(r)) - \mu\omega^2 r, \tag{5}$$

where for simplicity we suppose E and ε constant. Moreover $w(r)$ is assumed continuously differentiable. The problem is to find a function $w(r)$ minimizing

$$\int_{r_i}^{r_0} rw(r)\,dr \tag{6}$$

provided that the following conditions are satisfied:

$$w_m(r) \le w(r) \le w_M(r) \tag{7}$$

$$\sigma_r(r) \le \sigma_{rM}(r) \tag{8}$$

$$\sigma_t(r) \le \sigma_{tM}(r) \tag{9}$$

for all r, $r_i \le r \le r_0$.

$dw/dr(r)$ can be given on some subintervals of $[r_i, r_0]$.

The problem can be seen as an optimal-control problem. Define r as the independent variable; $w(r)$, $\sigma_r(r)$, $\sigma_t(r)$ as state variables; $\mathrm{d}w/\mathrm{d}r(r)$ as control variable. The problem is then to choose a function $\mathrm{d}w/\mathrm{d}r(r)$ which minimizes (6), such that conditions (7), (8), (9)—and, if there are any, conditions on $\mathrm{d}w/\mathrm{d}r(r)$—are satisfied, while the state variables satisfy the state equations and the limit conditions:

$$\frac{\mathrm{d}}{\mathrm{d}r}\, w(r) = \frac{\mathrm{d}w}{\mathrm{d}r}\,(r) \tag{10}$$

$$\frac{\mathrm{d}}{\mathrm{d}r}\, \sigma_r(r) = f_r\left(\frac{\mathrm{d}w}{\mathrm{d}r}\,(r),\, w(r),\, \sigma_r(r),\, \sigma_t(r),\, r\right) \tag{11}$$

$$\frac{\mathrm{d}}{\mathrm{d}r}\, \sigma_t(r) = f_t\left(\frac{\mathrm{d}w}{\mathrm{d}r}\,(r),\, w(r),\, \sigma_r(r),\, \sigma_t(r),\, r\right) \tag{12}$$

$$\sigma_r(r_i) = 0,\ \ \sigma_r(r_0) = \sigma_{r0},\ \ w(r_0) = w_0\,. \tag{13}$$

This optimal-control problem belongs to a class of problems rather difficult to solve by classical variational methods. The problem is state-constrained with inequalities and some of the state equations are nonlinear in the state and control variables.

4. A Dynamic-Programming Solution

In its continuous form dynamic programming leads to Hamilton-Jacobi-Bellman theory. However the major advantage of the dynamic-programming approach is in its discrete form, which makes it convenient for numerical solution. But the applicability of the standard algorithm is limited to relatively simple cases owing to large computer requirements. The most severe restriction is the amount of high-speed storage required. Another difficulty is the amount of computing time required to obtain the solution. For those reasons a problem with three or four state variables and one or two control variables seems to be the limit in practice. This is precisely the case of the disk problem.

Let us redefine the optimal-control problem in discrete form. Be

t the independent or stage variable (radius)

$$t_0 \le t \le t_m,$$

x (x_1, x_2, x_3) the state variables
(thickness, radial stress, tangential stress),
u the control variable (derivative of thickness).

The independent variable t is quantified in values indexed by the discrete sequence $i = 0, 1, \ldots, m$. The state and control variables corresponding to t_i are referred to as x^i and u^i.

The performance criterion (weight of the disk) is

(14) $$I = \sum_{i=0}^{m-1} F(x^i, u^i, t_i)\Delta_i, \text{ where}$$

(15) $$\Delta_i = t_{i+1} - t_i.$$

The state or dynamic equations, describing how the state variables change between t_{i+1} and t_i, are functions of x^i, u^i and i:

(16) $$x^{i+1} = f_i(x^i, u^i), \quad i = 0, \ldots, m-1.$$

They correspond to equations (10), (12) and (13) in their discrete form. The state constraints are

(17) $$x^i \in X^i \subset E^3, \quad i = 0, \ldots, m;$$

they correspond to (7), (8) and (9).

The control constraints are

(18) $$u^i \in U^i \subset E^1.$$

An iterative functional equation is defined in the following way [1]:

(19) $$S_{m-k}(x^{m-k}) = \min u^{m-k} \in U^{m-k} \{F(x^{m-k}, u^{m-k}, t_{m-k})\Delta_{m-k} + S_{m-k+1}(f_{m-k}(x^{m-k}, u^{m-k}))\}$$

for $k = 1, \ldots, m$ with

(20) $$S_m = 0.$$

When minimizing in u^{m-k} the expression bracketed in (19) for a given $x^{m-k} \in X^{m-k}$ constraint (17) has to be satisfied for $x^{m-k+1} = f(x^{m-k}, u^{m-k})$. The functional equation is computed backwards, $S_0(x^0)$ giving the optimal value of (14) for $x^0 \in X^0$.

Two problems arise when evaluating the functional equation at each stage t_i, the quantification of state and control variables, and the choice of state equations.

In order to apply the dynamic-programming computational procedure, there must be a finite number of admissible states and admissible controls.

This requirement is usually met by quantifying these variables. At each stage, the functional equation is then computed and tabulated over a grid of values of the state variables. Since constraints reduce the range of values of the state and control variables, a more constrained problem will require less computational effort. This is a valuable property of dynamic programming.

The state equation (16) is generally obtained by replacing in

$$\frac{dx}{dt} = f(x, u, t) \tag{21}$$

the derivative by approximate expressions such as

$$x^{i+1} - x^i = f(x^i, u^i, t_i)\Delta_i \tag{22}$$

$$x^{i+1} - x^i = \tfrac{1}{2}[f(x^i, u^i, t_i) + f(x^i + f(x^i, u^i, t_i)\Delta_i, u^i, t_i)]\Delta_i \tag{23}$$

or

$$x^{i+1} - x^i = \tfrac{1}{2}[f(x^i, u^i, t_i) + f(x^{i+1}, u^i, t_{i+1})]\Delta_i. \tag{24}$$

The choice of the increment in the quantification of the variables and the choice of an approximate expression for (21) influence accuracy. Since, when computing S_{m-k} for a given x^{m-k}, one has to interpolate in a table a value of S_{m-k+1}, the accuracy can be improved by decreasing the state increment, but in the meantime high-speed storage and computing time increase with the product of the number of levels of the different state variables. Another way of improving accuracy is to choose a good approximation for (21). Approximation (22) is rather crude and turned out to be the worst choice in our problem. Approximations of higher order such as (23) and (24) are better but increase computing time. Furthermore for stability considerations related to the choice of Δ_i, $i = 0, \ldots, m-1$, (23) and (24) are greatly preferable to (22).

Accuracy depends upon the choice of Δ_i, $i = 0, \ldots, m-1$ as well. The smaller the Δ_i, the more accurate the solution obtained. However, computing time increases linearly with the number of values of the stage variable.

5. A Numerical Example

Before presenting some numerical results, we show how we meet the problems arising when we use dynamic programming to solve the disk problem.

The quantification of the stage or independent variables corresponds to the choice of a particular sequence of radii r_i, $i = 0, \ldots, m$, where some of them are given by the geometrical constraints. For each stage i the corresponding set of admissible controls is given by the admissible sets of values for the thicknesses w_i and w_{i+1}, since the control variable corresponds to the slope of the profile. The profile obtained at the end of the optimization is piecewise linear.

When quantifying the state variables corresponding to stresses, one has to take an increment between 0.5% and 1% of their admissible range in order to get sufficient accuracy in their computation. Typical values for radial and tangential range, respectively, from 0 up to 60 hbar and from 0 to 75 hbar (1 hbar $\simeq$ 1.02 kg/mm^2). The grid of values for those variables can have between 4,500 and 18,000 points. Those numbers have to be multiplied by the number of levels of the first state variable, thickness, in order to obtain the number of storage locations at each stage. Since we want to keep the storage requirements at a reasonable level, we use a relatively small number of levels for the thickness. The optimization is worked out in several runs of the computer program.

For each run we use a more refined increment than in the previous one, and intervals centered around the values of the thickness obtained previously. For the same reason, we reduce the storage requirements by bounding artificially the range of values of the stresses. Since the true range is determined by the constraints of the problem, a solution where a stress is at an artificial bound is not optimal. A new run where this bound is changed is necessary.

Instead of using approximations such as (22), (23) or (24) for the differential equations (11) and (12), we use an idea which is at the basis of some programs devised to compute stresses of a disk of any shape. Since for disks of constant shape stresses are known analytically, one can use this solution when computing stresses in $r+dr$ knowing the stresses in r (fig. 2).

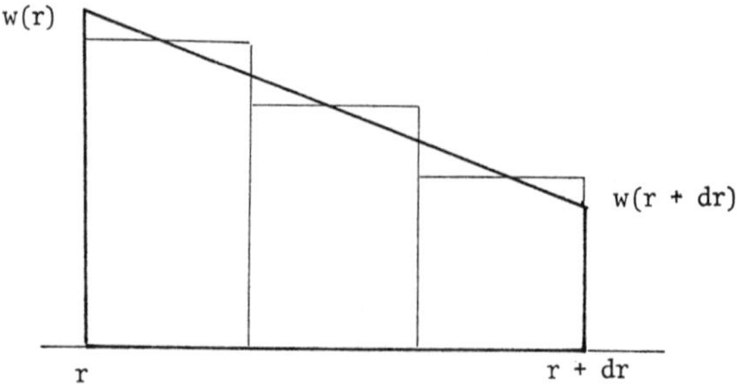

Fig. 2

This can be done in one step when the slope of the profile between r and $r+\mathrm{d}r$ is zero; otherwise the linear function $w(r)$ is approximated between r and $r+\mathrm{d}r$ by a step function with a number of steps proportional to the slope. Few iterations are needed to compute the stresses. This is the major advantage of the method. By comparison approximations such as (22), (23) and (24) would require many more steps independently of the slope of the profile.

By way of comparison we describe the typical profile of a disk. This profile is somewhat simplified; it consists of five parts. In the first part near the hub and in the fifth near the rim the profile is constant. In the middle part the profile corresponds to the shape of a uniform-strength disk. (Uniform strength means that σ_r and σ_t are constant throughout the disk.) This curve, known analytically, is tangentially connected by means of two parabolae to the two constant parts of the profile.

The data corresponding to this example are the following:

$\mu \;= 8{,}200$ kg/m^3
$\varepsilon \;= 14\cdot 10^{-6}$
$E \;= 18{,}525$ hbar
$N \;= 60\omega/2\pi = 22{,}000\ T/\mathrm{min}$
$r_i \;= 41$ mm
$r_0 \;= 131$ mm
$\sigma_{r0} = 18.50$ hbar.

The constraints of the optimization are the following:

Geometrical Constraints

— Maximum half-thickness 34 mm.
— Minimum half-thickness 4 mm.
— Between radii $r_0 = 131$ mm and $r_j = 127$ mm the half thickness has to be constant.
— The profile has to be constant near the hub.

Behavioral Constraints

$$\sigma_t(r_i) \leq 72 \text{ hbar}, \quad \sigma_t(r_0) \leq 47{,}5 \text{ hbar},$$
$$\sigma_r(r) \leq 55 \text{ hbar} \quad \text{for} \quad r_i \leq r \leq r_0.$$

In order to keep the storage and computing-time requirements at a reasonable level, we choose a sequence of 11 radii and we limit to 20,000 the number of possible states at each stage of the dynamic program. For the first run we use for the state variable intervals of values centered around the value of a feasible profile. So we are sure there should exist at least one feasible design.

The result of the first run is given in fig. 3. The computing-time and storage requirements corresponding to this example were respectively 7 min 45 sec.

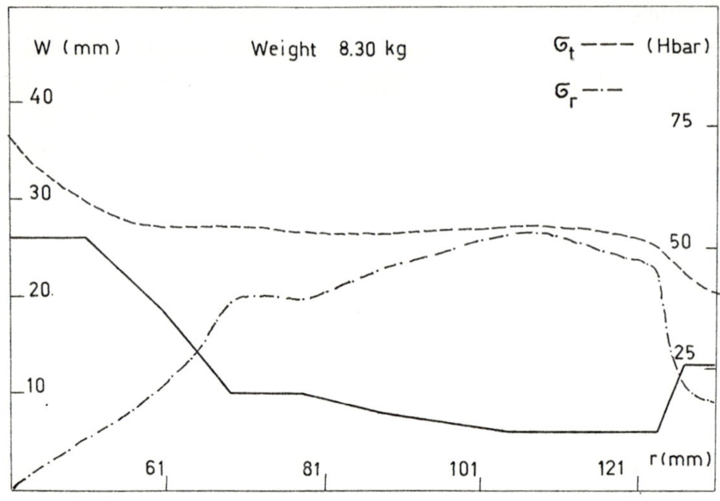

Fig. 3

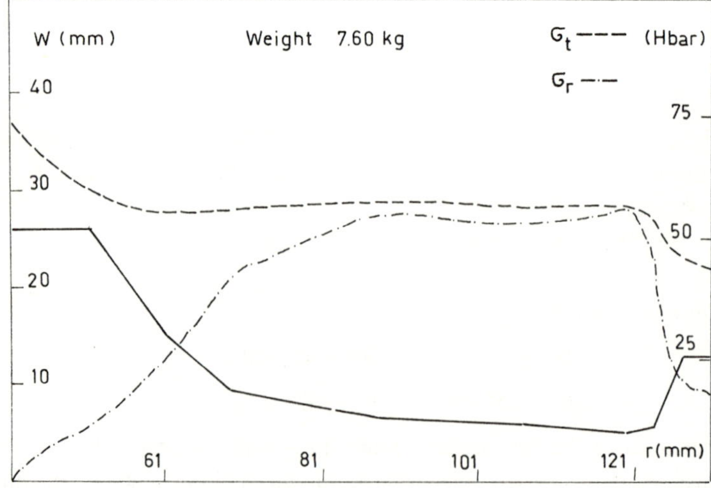

Fig. 4

and 270 K on IBM 370/155. The weight improvement is about 12% over the typical profile. Since at radius $r = 61$ mm an artificial lower bound of the thickness was attained, a second run of the program with more refined increments was needed. Fig. 4 gives the results. The total weight improvement over the typical profile is 18.7%.

It is worthwhile to observe that in spite of the different approximations and quantifications, for only a few values of the stresses are the constraints violated. Figures 3 and 4 give the true values for the stresses which are slightly different from the value given by the dynamic program.

6. Conclusions

In spite of its storage and computing-time requirements, dynamic programming is an interesting tool when designing turbine disks. An operational program should allow more than 50,000 possible states and use a greater number of stages than in our example. That means storage requirements of about 500 K and computing time of about half an hour. Such computing time is still comparable to the ones quoted in [3].

References

1. BELLMAN, R. and DREYFUS, S. *Applied Dynamic Programming*. Prentice Hall, New Jersey, 1971.
2. CHERN, J. M. and PRAGER, W. Optimal design of rotating disks for given radial displacement of edges. *JOTA*, **6**: 161–170, 1970.
3. DE SILVA, B. M. E. The application of nonlinear programming to the automated minimum weight design of rotating disks. FLETCHER, R. E., ed. *Optimization*. Academic Press, New York, 1969.
4. DI STEFANO, N. Dynamic programming and the optimum design of rotating disks. *JOTA*, **10**: 109–128, 1972.
5. NIGOGHOSSIAN, J. P. Note on "The optimization of Disks in Jet Engines", presented at the North Atlantic Science Council Advanced Summer School on Optimization in Technological Design, Louvain, July 1971.

CHAPTER 17

RESPONSIVE PRICING IN QUEUING SYSTEMS[1]

MAURICE G. MARCHAND

Summary

The purpose of this chapter is to investigate how the price imposed by a facility subject to queuing phenomena should fluctuate with queue size in the simple case where service times are exponentially distributed and no priority is introduced. It is assumed that customers are continuously informed of both current price and state of congestion. It is shown that, under plausible assumptions, efficiency requires the price to be an increasing function of the number of units in the system.

1. Introduction

Consider a computer center whose users are continuously kept aware of its degree of congestion, i.e. the time it will take to process all jobs present or, more realistically, the distribution of this time. When a user has to make up his mind about having some of his jobs executed, he takes into account how long he will have to wait for their completion and how much he will be charged by the computer center. Since his decision depends on the congestion of the computer center, the arrival rate of jobs varies with congestion.

As is well-known, self-regulation of the queue without pricing does not effect an efficient allocation of resources[2]. The objective of the paper is to investigate how such facility should charge its services in the simple case where service times are exponentially distributed and no priority is introduced. Of special interest is to study how the price should fluctuate with the state of the system[3], which is here simply defined as the number of jobs

[1] Research supported by the Ford Foundation under contract No. 69–720.

[2] For instance, see Naor (1969).

[3] The same problem is investigated by Vickrey (1971) in relation to the pricing of public utilities.

in the system. In particular, is the price a monotonic function of the number of jobs present? On intuitive grounds, it seems plausible to conjecture that the larger the number of jobs present at the arrival of a job, the higher should be the price charged. One can indeed observe that the expected number of jobs whose completion will be delayed by the incoming job before the facility gets idle is increasing with the number of jobs present at its arrival. We shall prove in section 3 this conjecture to be true under plausible assumptions.

The problem described above will be formulated by means of a general equilibrium model in which the price set by the facility is conditional on the state of the system[4]. It should however be stressed that the state-theoretic approach used here is somewhat untraditional because events will be viewed as repetitive. This will be apparent in section 3 whose developments will require to interpret the steady-state probabilities as the long-run proportions of time during which each state obtains.

The framework of this paper fits well the case where a stream of potential jobs is generated according to a Poisson process and the urgency and satisfaction associated with the processing of potential jobs vary at random from job to job. On the basis of the state of the system and the conditional price, the user takes for each of his jobs the *irrevocable decision* either to join the queue at once or to give up the idea of processing it. This rules out the opportunity to withdraw a job from the queue in order to recontract at a lower price if congestion happens to diminish. Moreover, it rules out the possibility of *speculating on the future state* of the system and to wait for a state of lower congestion before submitting a job, which could be very reasonable for jobs where urgency of completion is low.

2. The Unconstrained Case

We consider an economy where consumers derive satisfaction from the processing of jobs they submit to a one-server facility and from the consumption of other commodities which we summarize by a composite commodity.

The manager of the facility concludes with each of its users a contract

[4] While working on the present paper, the author received a report by David W. Low, "Optimal Dynamic Operating Policies for an M/M/s Queue with Variable Arrival Rate", I.B.M. Los Angeles Scientific Center Report G320–2654, in which the same kind of problem is formulated as a Semi-Markov Decision Process.

where the user stipulates the arrival rate of his jobs conditional on the state of the system. Because of the assumption of exponential service time that will be made, the state of the system is entirely defined by the number of jobs which are queuing or being processed. As the queue capacity is assumed to be unlimited, this number can take any positive integer value. It is denoted by n. We use λ_n^i to represent the arrival rate conditional on state n that is stipulated in user i's contract. More precisely, we suppose that when state n prevails and provided no other event, i.e. a departure or another arrival, occurs meanwhile, the waiting time until the next arrival of a job coming from user i is distributed as an exponential random variable whose mean is $(\lambda_n^i)^{-1}$. If we put side by side all intervals over which there are n jobs in the system, the arrival pattern of user i's jobs would look like a Poisson process with rate λ_n^i.

The processing times of the job are assumed to be identically and independently distributed and to be exponential random variables whose mean is denoted by μ^{-1}. Therefore, if a job arrives while the system is in state n, its waiting time in the system will have the Erlangian distribution with $(n+1)$ stages and mean equal to $(n+1)\mu^{-1}$.

We assume that each individual behaves according to the von Neumann-Morgenstern postulates and that his expected utility function can be written as

$$V^i = \sum_{n=0}^{\infty} P_n U_n^i(x^i, \lambda_n^i, \mu), \qquad i = 1, 2, \ldots, I, \tag{1}$$

where P_n denotes the steady-state probability of state n and x^i his consumption of the composite commodity. Since the waiting time of a job is a random variable, it should be clear that the utility function conditional on state n, that is U_n^i, is itself an expected utility function. We make the following assumptions concerning the utility functions:

A1: the conditional utility functions U_n^i are concave and twice continuously differentiable;

A2: no point of satiation is reached with respect to the composite commodity;

A3: $\partial U_n^i/\partial \lambda_n^i$ is strictly decreasing with n;

A4: $U_n^i(x^i, 0, \mu) = U_m^i(x^i, 0, \mu), \quad n, m = 0, 1, 2, \ldots$

The first two assumptions are classical, and the third one can be justified on intuitive grounds. Indeed, one expects that a reduction in the quality of service due to the presence of a larger number of jobs will lessen the amount users would be willing to pay for having their marginal jobs executed. As for the last assumption, it simply means that consumers would not care about the state of the facility if they dispensed with its services.

The consumer maximizes V^i with respect to x^i and $\lambda_n^i (n=0, 1, 2, \ldots)$ subject to his budget constraint and some nonnegativity conditions

$$(2) \qquad x^i + \sum_{n=0}^{\infty} \lambda_n^i q_n = y^i, \qquad i = 1, \ldots, I$$

$$x^i \geq 0, \qquad \lambda_n^i \geq 0,$$

where y^i denotes his income and q_n the price charged per unit of the n-state arrival rate. The price of the composite commodity is unity because it is used as numeraire.

Assuming once for all that $x_i > 0$ at the optimum, we obtain the following first-order conditions for a maximum:

$$(3) \qquad \begin{aligned} P_n \frac{\partial U_n^i}{\partial \lambda_n^i} \left(\frac{\partial V^i}{\partial x^i} \right)^{-1} &\leq q_n, \qquad \lambda_n^i = 0, \\ P_n \frac{\partial U_n^i}{\partial \lambda_n^i} \left(\frac{\partial V^i}{\partial x^i} \right)^{-1} &= q_n, \qquad \lambda_n^i > 0. \end{aligned}$$

Under the assumption that the arrival patterns of consumers are independent of each other, the steady-state probabilities of the different states are given by the following formulas[5]:

$$(4) \qquad P_n = P_0 \mu^{-n} \prod_{m=0}^{n-1} \lambda_m, \qquad n = 1, 2, \ldots$$

$$(5) \qquad \sum_{m=0}^{\infty} P_m = 1,$$

in which[6]

$$(6) \qquad \lambda_n = \sum_i \lambda_n^i, \qquad n = 0, 1, 2, \ldots$$

[5] See Cox and Smith (1960), p. 44. To apply these formulas, the service time as well as the time to the next arrival must be distributed as exponential random variables with means respectively equal to μ^{-1} and λ_n^{-1} if state n prevails.

[6] If W_n^i is an exponential random variable with parameter λ_n^i and W_n^i is independent of $W_n^j (i \neq j, i, j = 1, \ldots, I)$, then $W_n = \min(W_n^1, \ldots, W_n^I)^n$ is distributed as an exponential random variable with parameter λ_n given by (6).

Finally, the service rate of the facility can be improved at the expense of a reduction in the production of the composite commodity. Denoting this production by x we have

$$x = f(\mu). \tag{7}$$

Now, to find a Pareto optimum, we shall, according to chapter 1, section 2.2, maximize a weighted sum of the expected-utility functions (with arbitrary positive weights) subject to conditions (4)–(7) and the nonnegativity conditions $\lambda_n^i \geq 0$. The generalized Lagrangean of this problem is written as

$$\sum_i \beta^i V^i + \sum_{n=0}^{\infty} \gamma_n(\lambda_n - \sum_i \lambda_n^i) + \sum_{n=1}^{\infty} \pi_n(P_n - P_0\mu^{-n} \prod_{m=0}^{n-1} \lambda_m) + \tag{8}$$
$$+ \pi_0(\sum_{n=0}^{\infty} P_n - 1) + \phi[f(\mu) - \sum_i x^i].$$

After differentiation of the Lagrangean and some straightforward manipulations, we obtain the following first-order conditions for a Pareto optimum

$$P_n \frac{\partial U_n^i}{\partial \lambda_n^i}\left(\frac{\partial V^i}{\partial x^i}\right)^{-1} \leq \gamma_n \phi^{-1}, \qquad \lambda_n^i = 0,$$
$$P_n \frac{\partial U_n^i}{\partial \lambda_n^i}\left(\frac{\partial V^i}{\partial x^i}\right)^{-1} = \gamma_n \phi^{-1}, \qquad \lambda_n^i > 0, \tag{9}$$

$$\pi_n \phi^{-1} = \sum_i (\sum_{m=0}^{\infty} P_m U_m^i - U_n^i)\left(\frac{\partial V^i}{\partial x^i}\right)^{-1}, \quad n = 1, 2, \dots \tag{10}$$

$$\lambda_n \gamma_n = \sum_{m=n+1}^{\infty} \pi_m P_m, \qquad n = 0, 1, 2, \dots \tag{11}$$

$$-\frac{\partial f}{\partial \mu} = \sum_i \frac{\partial V^i}{\partial \mu}\left(\frac{\partial V^i}{\partial x^i}\right)^{-1} + \sum_{n=1}^{\infty} nP_n\mu^{-1}(\pi_n\phi^{-1}). \tag{12}$$

Comparing conditions (9) with conditions (3) which were found for the consumer optimum, we conclude that $\gamma_n\phi^{-1}$ can be interpreted as the price q_n

$$\gamma_n \phi^{-1} = q_n, \qquad n = 0, 1, 2, \dots \tag{13}$$

Substituting π_n from (10) into (11), and γ_n from (11) into (13), after some further manipulations, the first-order conditions for efficient prices can be shown to be equivalent to

$$(14) \qquad q_n = \sum_i \left(\frac{\partial V^i}{\partial x^i}\right)^{-1} \left[- \sum_{m=0}^{\infty} U_m^i \frac{\partial P_m}{\partial \lambda_n} \right], \qquad n = 0, 1, 2, \ldots$$

which can be given a straightforward interpretation in terms of welfare. Conditional on state n, the arrival rate should be increased up to the point where the price per unit of arrival rate equals the amount users of the facility would be willing to pay in order to prevent the steady-state probabilities from changing by $\partial P_m/\partial\lambda_n (m = 0, 1, 2 \ldots)$.

Likewise, the first-order condition for optimal service rate can be written from (12) as

$$(15) \qquad -\frac{\mathrm{d}f}{\mathrm{d}\mu} = \sum_i \left(\frac{\partial V^i}{\partial x^i}\right)^{-1} \left[\sum_{n=0}^{\infty} \left(P_n \frac{\partial U_n^i}{\partial \mu} + \frac{\partial P_n}{\partial \mu} U_n^i \right) \right].$$

This condition states that the service rate of the facility should be pushed up to the point where its marginal cost equals the amount users would be willing to pay in order to get both those improvements in the conditional waiting time distributions that result from the increase in μ (first term in parentheses on the right-hand side) and the changes $\partial P_n/\partial\mu$ in the steady-state probabilities (second term).

Needless to say, it would be very difficult in practice to verify whether conditions (14) and (15) are satisfied, due to the fact that the same steady-state probabilities and waiting-time distributions prevail for every user. In this kind of situation, one should expect to have difficulties in making everyone reveal his true willingness to pay.

3. Proof of the Monotonicity Property

The price we shall be concerned with in this section is not q_n itself. This price has indeed no time dimension since it represents what the facility charges per unit of arrival rate for the whole duration of the intervals over which there are n jobs in the system. The price for which we want to prove monotonicity should be defined per unit of time as well as per unit of arrival rate. Since P_n represents in the long run the proportion of time during which there are n jobs in the system, the price we are looking for can be seen to be proportional to the ratio q_n/P_n. It is this ratio that will be shown to be increasing with n.

To simplify the proof, we shall take the case where there is only one individual (and delete the index i).

Let

$$N = \min(n : \lambda_n = 0).$$

The existence of this number is granted by the third of the assumptions on the utility functions. From (11), we obtain the following recurrence relationships:

$$\lambda_{N-1}\gamma_{N-1} = \pi_N P_N, \tag{16}$$

$$\lambda_n \gamma_n = \pi_{n+1} P_{n+1} + \lambda_{n+1}\gamma_{n+1}, \qquad n = 0, 1, \ldots, N-2. \tag{17}$$

Now, we first eliminate the γ's in the above relations by means of (13) for those appearing on the left-hand side and of (9) for those on the other side. As for the π's, they are substituted from (10). Finally, some further manipulations in which we use the relation $P_n = \lambda_{n-1} P_{n-1} \mu^{-1}$ yield for the one-consumer case

$$\mu \frac{q_{N-1}}{P_{N-1}} = A - \left(\frac{\partial V}{\partial x}\right)^{-1} U_N, \tag{18}$$

$$\mu \frac{q_n}{P_n} = A - \left(\frac{\partial V}{\partial x}\right)^{-1} \left(U_{n+1} - \lambda_{n+1} \frac{\partial U_{n+1}}{\partial \lambda_{n+1}}\right), \qquad n = 0, 1, \ldots, N-2 \tag{19}$$

where

$$A = \left(\frac{\partial V}{\partial x}\right)^{-1} \left(\sum_{m=0}^{N} P_m U_m\right).$$

To complete the proof, we shall first prove the inequality

$$\frac{q_{N-1}}{P_{N-1}} > \frac{q_{N-2}}{P_{N-2}}. \tag{20}$$

From (18) and (19), it is equivalent to prove

$$U_{N-1} - U_N > \lambda_{N-1} \frac{\partial U_{N-1}}{\partial \lambda_{N-1}}. \tag{21}$$

The proof of the last inequality follows from the first and fourth assumptions on the utility functions.

Next, we shall prove

$$\frac{q_{N-2}}{P_{N-2}} > \frac{q_{N-3}}{P_{N-3}}. \tag{22}$$

To do so, we infer from the appendix that under the assumptions on utility functions, inequality (20) implies that $\lambda_{N-1} < \lambda_{N-2}$. As shown again in the appendix, the following relationship then follows from the same assumptions

$$U_{N-2} - \lambda_{N-2} \frac{\partial U_{N-2}}{\partial \lambda_{N-2}} > U_{N-1} - \lambda_{N-1} \frac{\partial U_{N-1}}{\partial \lambda_{N-1}}. \tag{23}$$

In view of (19), the last inequality is equivalent to (22).

Repeating step by step the same reasoning yields

$$\frac{q_n}{P_n} > \frac{q_{n-1}}{P_{n-1}}$$

for every n ($= 0, 1, \ldots, N-1$), which ends the proof.

4. Concluding remarks

It was mentioned at the end of the introductory section that the model presented in this paper did not allow recontracting at a lower price and waiting for a lower price, which causes obvious limitations in its usefulness. Taking account of these possibilities is, however, likely to make the model very intricate. If the facility is highly congested at some time, the users would indeed be induced to defer part of their demand to the near future in the hope of avoiding too high a price. This means that one should take into consideration that the arrival rate of jobs at any one time would depend upon the past history of the facility. Moreover, to deal properly with this kind of situation, one should probably introduce both spot and futures markets for the services of the facility.

To verify whether current prices and service rates of a facility satisfy the first-order conditions for an efficient allocation of resources, its users must correctly reveal their willingness to pay, which might be impossible. In another context[7], it was shown that in some circumstances, the introduction of priority levels allows to infer from the priority levels chosen by the users information about their willingness to pay for reductions in waiting

[7] See Marchand (1971).

times. It would be interesting to investigate whether the same property carries over to the present case.

Appendix

We shall prove the following theorem:

Let $g(x, y)$ be a function defined for every $x \geq 0$ and $y \in \{0, 1, 2, \ldots\}$. Assume that

a1: g is twice differentiable and concave with respect to x for every $x \geq 0$;

a2: $g_x(x, y_0) > y_x(x, y_1)$ for every y_0 and y_1 such that $y_1 > y_0$.

If x_0 and x_1 are chosen such that $g_x(x_0, y_0) < g_x(x_1, y_1)$ with $y_0 < y_1$, then the following inequality holds:

$$g(x_0, y_0) - g(0, y_0) - x_0 g_x(x_0, y_0) > g(x_1, y_1) - g(0, y_1) - x_1 g_x(x_1, y_1).$$

Proof.

A. We first show that $x_0 > x_1$. Let x_2 be such that $g_x(x_2, y_1) = g_x(x_0, y_0)$. By the assumption made on the choice of x_0 and x_1, we have $g_x(x_2, y_1) < g_x(x_1, y_1)$, which implies $x_2 > x_1$ by the first assumption. On the other hand, we have $g_x(x_2, y_0) > g_x(x_2, y_1) = g_x(x_0, y_0)$ by the second assumption, which implies $x_0 > x_2$ by the first assumption. So $x_0 > x_2 > x_1$.

B. The desired inequality is equivalent to

$$\int_0^{x_0} [g_x(u, y_0) - g_x(x_0, y_0)]\,\mathrm{d}x > \int_0^{x_1} [g_x(u, y_1) - g_x(x_1, y_1)]\,\mathrm{d}x.$$

Since $x_0 > x_1$, and $g_x(u, y_0) - g_x(x_0, y_0) > g_x(u, y_1) - g_x(x_1, y_1) > 0$ by the first and second assumptions, this inequality holds. QED.

References

1. Cox, D. R. and Smith, W. L. *Queues.* Methuen, London, 1960.
2. Marchand, M. G. Priority pricing. *CORE D.P. No.* 7124, 1971.
3. Naor, P. On the regulation of queue size by levying tolls. *Econometrica*, **37**: 15–24, 1969.
4. Vickrey, W. Responsive pricing of public utility services. *The Bell Journal of Economics and Management Science*, **2**: 337–346, 1971.

CHAPTER 18

ON MARKOV SUCCESS CHAINS[1]

JOZEF L. TEUGELS

Summary

This paper attempts to deal with a few of the basic characteristics of a success chain consisting of a finite or a denumerable number of states. In every state there is a possibility of failure, tie, or success with probabilities depending on the state. Apart from classification properties we treat geometric ergodicity and also quasi-stationarity in the presence of a terminal absorbing state.

1. Definitions and Notations

Consider a stationary discrete-time Markov chain on the state space $I=\{0, 1, 2, \ldots\}$ with one-step transition matrix $P=(P_{ij})$, $i, j\in I$, where for $i>0$

$$P_{ij}=\begin{cases} q_i & \text{if} \quad j=i+1 \\ r_i & \text{if} \quad j=i \\ p_i & \text{if} \quad j=0 \end{cases}$$

and $P_{00}=p_0$, $P_{01}=q_0$. We assume that $p_i+r_i+q_i=1$ for $i>0$ and $p_0+q_0=1$.

The above Markov chain occurs in situations where one has to pass through a number of different trials, each trial having its own chance of failure, tie, or success. The trials are mutually independent but the entire cycle of trials has to be repeated as soon as one runs into a failure. Examples are assembly lines, exams, computer-aided instruction, mail delivery, the slabbing department in a steelmill, etc.

[1] We would like to thank Dr. J. de Smit and N. Veraverbeke for valuable criticism and suggestions.

To make the model realistic we assume irreducibility of the chain: this is satisfied if $q_i > 0$ for all $i \in I$ and $p_i > 0$ for infinitely many $i \in I$. We also assume that the chain is aperiodic although this restriction can be avoided at the cost of notational clearness. A sufficient condition for aperiodicity is that

$$p_0 + \sum_{i=1}^{\infty} r_i > 0.$$

We introduce some notation:

$$u_{-1} = 1$$

$$u_n = q_0 \, q_1 \cdots q_n, \quad n \geq 0;$$

(1) $$\underline{r} = \inf_{i \in I} r_i, \qquad \bar{r} = \sup_{i \in I} r_i,$$

(2) $$\underline{q} = \inf_{i \in I} q_i, \qquad \bar{q} = \sup_{i \in I} q_i.$$

If for $i > 0$, $r_i = r$ where $0 \leq r < 1$ the success chain will be called *elementary*; if $r = 0$ in the elementary case, we call the chain the *basic chain.*

One encounters the basic chain in a number of textbooks on stochastic processes [2, 6, 8]. Especially [9] gives an extensive study of the basic chain in the general framework of potential theory.

2. Classification of States

Due to the irreducibility of the success chain we only need two quantities to classify the states. Let $\{X_m, m = 0, 1, 2, \ldots\}$ be the Markov chain. Put

$$f_{ij}^{(n)} = P\{X_n = j, X_m \neq j \quad \text{for} \quad 0 < m < n \mid X_0 = i\}$$

then $\{f_{ij}^{(n)}; n > 0\}$ is the first passage distribution from state i into state j. Its generating function will be denoted by $F_{ij}(z)$ while that of the n-th step transition probabilities is denoted by $P_{ij}(z) = \sum_{n=0}^{\infty} P_{ij}^{(n)} z^n$ where

$$P_{ij}^{(n)} = P\{X_n = j \mid X_0 = i\}.$$

Because of the special nature of state 0 we use $F_{00}(z)$ in classifying states. We recall that the chain is transient if $F_{00}(1) < 1$, recurrent if $F_{00}(1) = 1$. A recurrent chain is null recurrent if $F'_{00}(1) = \infty$, is positive (recurrent) if $F'_{00}(1) < \infty$.

LEMMA 1.

(i) $F_{00}(z) = p_0 z + z \sum_{m=1}^{\infty} u_{m-1} p_m z^m \prod_{j=1}^{m} (1-r_j z)^{-1}$;

(ii) $\dfrac{1-F_{00}(z)}{1-z} = 1 + \sum_{m=1}^{\infty} u_{m-1} z^m \prod_{j=1}^{m} (1-r_j z)^{-1}$.

Proof.

(i) Let Y be the first return time from state 0 to state 0. The probability that state m is the last state visited prior to a first return to state 0 is $q_0 q_1 \dots q_{m-1} p_m = u_{m-1} p_m$.
Denote by N_i the number of returns to state i during the return time Y. Since the N_i are independent we have

$$F_{00}(z) = E\{z^Y\} = p_0 z + \sum_{m=1}^{\infty} u_{m-1} p_m E\{z^{N_1+\dots+N_m+m+1}\}$$

$$= p_0 z + z \sum_{m=1}^{\infty} u_{m-1} p_m \prod_{j=1}^{m} E\{z^{N_j}\} z^m .$$

Now N_j is geometrically distributed with parameter r_j and hence $E\{z^{N_j}\} = (1-r_j z)^{-1}$ from which (i) follows.

(ii) Replacing $p_m z$ by $(z-1)+(1-r_m z)-q_m z$ for $m \geq 1$ in (i) yields (ii) after some algebra. QED.

We obtain criteria for classification from the above expressions. Recurrence is decided by (i), positive recurrence by (ii).

THEOREM 1. The success chain is

(i) transient if $\lim_{n\to\infty} u_n \prod_{j=1}^{n} (1-r_j)^{-1} \neq 0$;

(ii) null recurrent if $\begin{cases} \lim_{n\to\infty} u_n \prod_{j=1}^{n} (1-r_j)^{-1} = 0 \\ \sum_{m=1}^{\infty} u_{m-1} \prod_{j=1}^{m} (1-r_j)^{-1} = \infty ; \end{cases}$

(iii) positive if $\sum_{m=1}^{\infty} u_{m-1} \prod_{j=1}^{m} (1-r_j)^{-1} < \infty$.

It is not difficult to replace the condition for transience by a simpler one. Unfortunately positive recurrence cannot be handled as easily.

COROLLARY 1. The success chain is transient if

$$\sum_{n=1}^{\infty} p_n(p_n+q_n)^{-1}<\infty .$$

Proof. This follows along the same lines as for the case $r_i=0$. We refer to [8, p. 52] or [12, p. 231].

For the recurrent case define recursively

$$n_0 = \inf\{n\geq 0: p_n>0\},$$

$$n_{m+1} = \inf\{n>n_m: p_n>0\} \quad \text{for} \quad m\geq 0 .$$

Corollary 1 implies that in the recurrent case

$$\sum_{m=0}^{\infty} p_{n_m}(p_{n_m}+q_{n_m})^{-1} = \infty . \tag{3}$$

An easy calculation reveals that

$$\sum_{k=1}^{\infty} u_{k-1} \prod_{j=1}^{k} (1-r_j)^{-1} = \sum_{m=0}^{\infty} \frac{q_{n_1}}{1-r_{n_1}} \cdot \frac{q_{n_2}}{1-r_{n_2}} \cdots \frac{q_{n_m}}{1-r_{n_m}} \sum_{j=n_m+1}^{n_{m+1}} (1-r_j)^{-1} .$$

This series converges (diverges) if the series

$$\sum_{m=0}^{\infty} (n_{m+1}-n_m) \frac{q_{n_1}}{1-r_{n_1}} \cdots \frac{q_{n_m}}{1-r_{n_m}}$$

converges (diverges), if $\bar{r}<1$ where $\bar{r}$ is defined by (1). Using the divergent series above as comparison series in Kummer's criterion [12, p. 320] we obtain

COROLLARY 2. Let the success chain be recurrent:

(a) if for m large enough

$$\frac{n_{m+1}-n_m}{n_{m+2}-n_{m+1}} \leq \frac{p_{n_m}}{p_{n_{m+1}}} q_{n_{m+1}}(p_{n_m}+q_{n_m})^{-1}$$

then the chain is null recurrent;

(b) if $\bar{r}<1$ and if for m large enough

$$\frac{n_{m+1}-n_m}{n_{m+2}-n_{m+1}} > \delta + \frac{p_{n_m}}{p_{n_{m+1}}} q_{n_{m+1}} (p_{n_m}+q_{n_m})^{-1}$$

for some $\delta>0$, then the chain is positive.

We remark that the conditions in (a) and (b) are easily checked in practice.

COROLLARY 3. In the positive case the stationary distribution of the Markov chain is given by

$$\lim_{n\to\infty} P_{ik}^{(n)} = \begin{cases} c \equiv \{1+\sum_{m=1}^{\infty} u_{m-1} \prod_{j=1}^{m} (1-r_j)^{-1}\}^{-1} & \text{if} \quad k=0 \\ cu_{k-1} \prod_{j=1}^{k} (1-r_j)^{-1} & \text{if} \quad k>0. \end{cases}$$

3. The Functions $F_{ij}(z)$

In this section we derive explicit expressions for the first recurrence-generating functions. As an amusing side result it will turn out that the recurrence distributions can be expressed in terms of that of state 0.

The following identity will help us in deriving $F_{ii}(z)$:

$$F_{ij}(z) = {}_kF_{ij}(z) + {}_jF_{ik}(z) F_{kj}(z), \qquad k \neq j.$$

Choose i, j and k among the two states 0 and $i \neq 0$. We obtain four auxiliary relations:

(4) $$F_{00}(z) = {}_iF_{00}(z) + {}_0F_{0i}(z) F_{i0}(z),$$

(5) $$F_{0i}(z)\{1-{}_iF_{00}(z)\} = {}_0F_{0i}(z),$$

(6) $$F_{i0}(z)\{1-{}_0F_{ii}(z)\} = {}_iF_{i0}(z),$$

(7) $$F_{ii}(z) = {}_0F_{ii}(z) + {}_iF_{i0}(z) F_{0i}(z).$$

Elimination of ${}_0F_{0i}(z)\, F_{i0}(z)$ and ${}_iF_{i0}(z)\, F_{0i}(z)$ yields a relation originally due to Vere-Jones [16]:

(8) $$\frac{1-F_{ii}(z)}{1-F_{00}(z)} = \frac{1-{}_0F_{ii}(z)}{1-{}_iF_{00}(z)}.$$

Returning to our success chain we immediately have

$$ {}_0F_{ii}(z) = r_i z, \qquad i>0; \tag{9} $$

by using the same method of proof as in lemma 1 (i) we also get

$$ {}_iF_{00}(z) = p_0 z + z \sum_{m=1}^{i-1} u_{m-1} p_m z^m \prod_{j=1}^{m} (1-r_j z)^{-1}, \tag{10} $$

illustrating that $F_{00}(z) = \lim_{i\to\infty} {}_iF_{00}(z)$.

THEOREM 2. For $n>1$ and $i \neq 0$

$$ P_{ii}^{(n)} - r_i P_{ii}^{(n-1)} = P_{00}^{(n)} - \sum_{k=1}^{\min(i,n)} f_{00}^{(k)} P_{00}^{(n-k)}. \tag{11} $$

Proof. Since $P_{ii}(z) = \{1-F_{ii}(z)\}^{-1}$, (8) and (9) imply

$$ (1-r_i z) P_{ii}(z) = [1 - {}_iF_{00}(z)]\, P_{00}(z). $$

But comparing lemma 1 (i) with (10) we also see that

$$ \begin{aligned} {}_if_{00}^{(k)} &= f_{00}^{(k)} \quad \text{if} \quad k \le i \\ &= 0 \quad \text{if} \quad k > i, \end{aligned} $$

and this proves (11). QED.

COROLLARY 4. For the basic chain and $n \ge i > 0$

$$ P_{ii}^{(n)} = P_{00}^{(n)} - \sum_{k=1}^{i} [u_{k-2} - u_{k-1}]\, P_{00}^{(n-k)}. $$

COROLLARY 5. If μ_{ii} is the mean recurrence time to state i, then

$$ u_{i-1}\mu_{ii} = (1-r_1)\dots(1-r_i)\mu_{00}. $$

Also the functions $F_{0i}(z)$ and $F_{i0}(z)$ are now explicitly known from (4) to (7) by using for example

$$ {}_0F_{0i}(z) = u_{i-1} z^i \prod_{j=1}^{i-1} (1-r_j z)^{-1}, \qquad i \neq 0. $$

4. Geometric Ergodicity

The formulas derived in the previous section enable us to deal with the concept of geometric ergodicity as a solidarity property of the states in a success chain. We rely on the next lemma which is due to Kendall [10] and Vere-Jones [16].

LEMMA 2. In an irreducible aperiodic stationary discrete-time Markov chain on I, the following conditions are equivalent:

(i) state 0 is geometrically ergodic, i.e. for some finite constant K_0 and some minimal real number ρ_0, $0 \leq \rho_0 < 1$, called the *decay parameter* of state 0, we have that for all $n \geq n_0$

$$|P_{00}^{(n)} - \lim_{n\to\infty} P_{00}^{(n)}| \leq K_0 \rho_0^n;$$

(ii) state i is geometrically ergodic with decay parameter ρ_i, $0 \leq \rho_i < 1$;

(iii) $F_{00}(z)$ is analytic in an open set containing the unit disc $\{z:|z| \leq 1\}$;

(iv) $(1-z)\,\{1-F_{00}(z)\}^{-1}$ is analytic in an open set containing the unit disc.

Moreover in the transient case $\rho_i = \rho_0$ for all $i \in I$; in the positive case

$$\sup_{i\in I} \rho_i < 1.$$

As a consequence of (8), (9) and (10) we obtain

LEMMA 3. On the positive and geometrically ergodic case, $\rho_i \leq \rho_0$ for all $i \in I$.

Proof. By lemma 2 (iv) and (8) we can write:

$$\frac{1-z}{1-F_{ii}(z)} = \frac{1-z}{1-F_{00}(z)} \cdot \frac{1-{}_iF_{00}(z)}{1-{}_0F_{ii}(z)}. \tag{12}$$

Let ρ_0^{-1} be the radius of convergence of $(1-z)\,\{1-F_{00}(z)\}^{-1}$ then $\rho_0^{-1} < \bar{r}^{-1}$ by lemma 1, (ii). Moreover $1-{}_iF_{00}(z)$ is analytic in $\{z:|z| < \bar{r}^{-1}\}$ while $1-{}_0F_{ii}(z) = 1-r_i z$ is zero free in this circle. Hence the radius of convergence of the left hand side of (12) is at least ρ_0^{-1}. QED.

From lemma 3 we infer that ρ_0 is "the" decay parameter of the success chain. Its actual value is determined either by the region of convergence of $F_{00}(z)$ or by the zero with smallest absolute value of $\{1-F_{00}(z)\}\,(1-z)^{-1}$.

In general it is hard to get information on that zero; the next theorem is only a partial result in this direction.

Put

$$\frac{1}{R} = \limsup_{n\to\infty} \sqrt[n]{u_{n-1}p_n}. \tag{13}$$

THEOREM 3.

(i) The transient chain is geometrically ergodic if $R+\bar{r}<1$; furthermore $1+\underline{r}R \leq \rho_0 R \leq 1+\bar{r}R$.

(ii) The elementary chain is positive and geometrically ergodic if $r+\bar{q}<1$; furthermore $r+\underline{q} \leq \rho_0 \leq r+\bar{q}$.

Proof.

(i) From lemma 1 (i) there results that $F_{00}(z)$ is the generating function on a nonnegative sequence and henceforth it has its first singularity on the positive real axis in view of the Vivanti-Pringsheim theorem [15, p. 145]. We can obviously restrict our attention to the interval $1 \leq x < 1/r$. For this range

$$p_0 x + \sum_{m=1}^{\infty} u_{m-1} p_m (x/1-\underline{r}x)^m \leq F_{00}(x) \leq p_0 x + \sum_{m=1}^{\infty} u_{m-1} p_m (x/1-\bar{r}x)^m,$$

and henceforth $F_{00}(x)$ converges if $x(1-\bar{r}x)^{-1}<R$ and diverges if $x(1-\underline{r}x)^{-1}>R$. This proves (i).

(ii) We use lemma 1 (ii) and $r_i = r$ to write

$$\frac{1-F_{00}(z)}{1-z} = 1 + \sum_{k=1}^{\infty} v_k z^k \equiv V(z)$$

where $v_k = \sum_{l=0}^{k-1} \binom{k-1}{l} u_l r^{k-l-1}$. Put $v_0 = 1$ and $V_n(z) = \sum_{k=0}^{n} v_k z^k$.

By a result of Kakeya [13, p. 88] all zeros of $V_n(z)$ are situated in the annulus

$$\left\{\max_{0\leq k<n} \frac{v_{k+1}}{v_k}\right\}^{-1} \leq |z| \leq \left\{\min_{0\leq k<n} \frac{v_{k+1}}{v_k}\right\}^{-1}.$$

But

$$\frac{v_{k+2}}{v_{k+1}} = r+(1-r)\left\{\sum_{l=0}^{k} u_{l+1}\binom{k}{l} r^{k-l}\right\}\left\{\sum_{l=0}^{k} u_l \binom{k}{l} r^{k-l}\right\}^{-1}$$

so that $r+q \leq (v_{k+2}/v_{k+1}) \leq r+\bar{q}$ for all $k \geq 0$. Henceforth $V_n(z)$ has all zeros in the annulus

$$\{r+\bar{q}\}^{-1} \leq |z| \leq \{r+q\}^{-1}.$$

By Hurwitz's theorem [1] $V(z)$ has no zeros in $|z|<(r+\bar{q})^{-1}$ since $V(0)=1$. Since $V(z)$ is clearly analytic in this circle, the same is true for the function $(1-z)\,\{1-F_{00}(z)\}^{-1}$. By lemma 2 (iv) the result follows if we can show that the chain is positive. But this follows from corollary 1 since $p_n(p_n+q_n)^{-1} \geq (1-r-\bar{q})\,(1-r)^{-1}>0$, and from the fact that $1/V(z)$ cannot converge outside the unit disc in the null recurrent case. This proves (ii).

COROLLARY 6. If in the elementary transient case $\lim_{n\to\infty} \sqrt[n]{p_n} \equiv p^{-1}$ then the chain is geometrically ergodic if and only if $p>1$ and $\rho_0 = r+(1-r)/p$.

Proof. Since $u_{n-1} \prod_{k=1}^{n-1} (1-r_k)^{-1} \to d>0$ and since $r_i = r$ we obtain that $1/R=(1-r)/p$. Using this in theorem 3 (i) we get the desired result. QED.

5. The Finite Case

We now turn to the case where state $N \geq 1$ is absorbing, i.e. the one-step transition matrix takes the form

$$P = \begin{bmatrix} p_0 & q_0 & 0 & \dots & 0 & 0 \\ p_1 & r_1 & q_1 & \dots & 0 & 0 \\ p_2 & 0 & r_2 & \dots & 0 & 0 \\ \cdot & \cdot & \cdot & \dots & \cdot & \cdot \\ p_{N-1} & 0 & 0 & \dots & r_{N-1} & q_{N-1} \\ 0 & 0 & 0 & \dots & 0 & 1 \end{bmatrix}.$$

We keep the assumption that $q_i>0$ for all $i \in I_N = \{0, 1, \dots, N-1\}$ and that $p_{N-1}>0$. Furthermore we assume that the finite chain is aperiodic.

The next lemma is due to Darroch-Seneta [3] and Seneta-Vere-Jones [14].

LEMMA 4. The irreducible Markov chain with transition matrix P restricted to I_N is called R_N-positive if the limits

$$L_{ij} = \lim_{n\to\infty} R_N^n P_{ij}^{(n)}$$

exist, are finite and positive. This happens if there exist nonnegative nonzero solutions to the two systems of equations

$$\text{(i)} \quad R_N \sum_{k=0}^{N-1} m_k P_{kj} = m_j \qquad j \in I_N$$

$$\text{(ii)} \quad R_N \sum_{k=0}^{N-1} P_{jk} x_k = x_j \qquad j \in I_N .$$

Moreover $L_{ij} = c x_i m_j$ where $c^{-1} = \sum_{k=0}^{N-1} x_k m_k$.

THEOREM 4. There exists a value $R_N > 1$ such that

$$\lim_{n\to\infty} R_N^n P_{ij}^{(n)} = L_{ij}$$

where $0 < L_{ij} < \infty$ for all $i, j \in I_N$.

Proof. The system (i) can be rewritten in the form

$$\begin{cases} R_N \sum_{k=0}^{N-1} m_k p_k = m_0 \\ R_N m_{j-1} q_{j-1} = m_j(1 - r_j R_N) \qquad j = 1, 2, \ldots, N-1 . \end{cases}$$

Using m_0 as a normalizing constant his system has a positive solution if R_N satisfies $f_N(R_N) = 1$ where

$$f_N(R) = p_0 R + \sum_{k=1}^{N-1} u_{k-1} p_k R^{k+1} \prod_{j=1}^{k} (1 - r_j R)^{-1} . \tag{14}$$

But as a function of $R > 0$, $f_N(R)$ is nonnegative and increasing for R increasing; as $R \to 1/r$ where $r = \max_{1 \leqslant k \leqslant N-1} r_k$, $f_N(R)$ tends to $+\infty$. Henceforth there will exist a unique value R_N such that $R_N > 1$ and for which $f_N(R_N) = 1$ if $f_N(1) < 1$. By using $p_k = 1 - r_k - q_k$ this can be rewritten as

$$f_N(1) = 1 - u_{N-1} \prod_{j=1}^{N-1} (1 - r_j)^{-1} < 1 .$$

With this R_N we obtain

(15) $$m_k = R_N^k u_{k-1} \prod_{j=1}^{k} (1-r_j R_N)^{-1} m_0 \qquad k=1,2,\ldots,N-1.$$

The second system has the form

$$\begin{cases} (1-R_N p_0)x_0 = R_N q_0 x_1 \\ (1-R_N r_k)x_k = R_N q_k x_{k+1} + R_N p_k x_0, & 1 \leq k \leq N-2 \\ (1-R_N r_{N-1})x_{N-1} = R_N p_{N-1} x_0. \end{cases}$$

Using x_0 now as normalizing constant this system leads again to the condition $f_N(R_N)=1$. The explicit solution is now

$$x_1 = R_N^{-1} u_0^{-1} \{1-R_N p_0\} x_0$$

$$x_k = R_N^{-k} u_{k-1}^{-1} \prod_{j=1}^{k-1} (1-R_N r_j) \times$$
$$\times \{1-p_0 R_N - \sum_{j=1}^{k-1} R_N^{j+1} u_{j-1} p_j \prod_{l=1}^{j} (1-r_l R_N)^{-1}\} x_0, \qquad 2 \leq k \leq N-1.$$

With the help of $f_N(R_N)=1$ we can rewrite this in the simpler form

(16) $$m_k x_k = (1-r_k R_N)^{-1} x_0 m_0 \sum_{j=k}^{N-1} R_N^{j+1} u_{j-1} p_0 \prod_{l=1}^{j} (1-r_l R_N)^{-1}, \qquad 1 \leq k \leq N-1.$$

From (15) and (16) we can easily obtain the L_{ij}. QED.

COROLLARY 7. For the finite success chain the strong ratio-limit theorem holds, i.e.

$$\lim_{n\to\infty} \frac{P_{ij}^{(n)}}{P_{kl}^{(n)}} = \frac{L_{ij}}{L_{kl}}.$$

COROLLARY 8. For the finite success chain the quasi-stationary distribution exists, i.e.

$$V_{ij} = \lim_{n\to\infty} P_{ij}^{(n)} \Big/ \sum_{k=0}^{N-1} P_{ik}^{(n)}$$

exists for every $i, j \in I_N$.

Proof. As defined in [3] the quasi-stationary distribution gives the limit as $n \to \infty$ of the probability that the Markov chain on I_N is in state j given that it started in i and is still not absorbed at the n-th transition. Using theorem 4 we find that $V_{ij}=m_j\{\sum_{k=0}^{N-1} m_k\}^{-1}$, independently of i. QED.

COROLLARY 9. As a function of N, R_N satisfies the inequalities

$$1<\frac{R_N-1}{R_{N+1}-1}\leq 1+R_N^{N+1}\left(\frac{p_N}{q_N}\right)\prod_{j=1}^{N}\left(\frac{1-r_j}{1-r_jR_N}\right). \tag{17}$$

Proof. From (14) $f_{N+1}(R_N)=f_N(R_N)+\delta_{N+1}=1+\delta_{N+1}$ where

$$\delta_{N+1}=R_N^{N+1}u_{N-1}p_N\prod_{j=1}^{N}(1-r_jR_N)^{-1}.$$

We draw the graphs of the functions $f_N(R)$ and $f_{N+1}(R)$. We join the points $[1, f_{N+1}(1)]$ by a straight line, which meets the horizontal line through the point (0,1) in the point $(t_{N+1}, 1)$. Henceforth

$$\{\delta_{N+1}+u_N\prod_{j=1}^{N}(1-r_j)^{-1}\}(t_{N+1}-1)=(R_N-1)u_N\prod_{j=1}^{N}(1-r_j)^{-1}.$$

But $f_{N+1}(R)$ is convex in the interval $1\leq R\leq R_N$ and hence $t_{N+1}\leq R_{N+1}$. This proves the right inequality of (17).

Since $\delta_{N+1}>0$ by the assumption that $p_N>0$, we also have $R_{N+1}<R_N$ from which the inequality on the left of (17) follows. QED.

6. Examples

1) *The basic chain with q_i non-decreasing*

Although the applicability of this example might be restricted, a few side results are extremely appealing.

One of the interesting features of this example results from a theorem of Szegö, referred to in Hardy [7, p. 68]. Consider state 0, then by lemma 1 (ii)

$$(1-z)P_{00}(z)=\{1+\sum_{n=1}^{\infty}u_{n-1}z^n\}^{-1}\equiv\{U(z)\}^{-1}.$$

But now $u_{n+1}u_{n-1}\geq u_n^2$ since the q_n are non-decreasing and hence $\alpha_n\equiv P_{00}^{(n)}-P_{00}^{(n+1)}\geq 0$ for $n=0, 1, 2, \ldots$.

The Vivanti-Pringsheim theorem applies again to the series with coefficients $\alpha_n\geq 0$, and hence $(1-z)P_{00}(z)$ converges for $|z|<R$ where R is the

first singularity of $1/U(z)$ on the positive real axis. Since $U(x)>0$ for $x>0$, $R^{-1}=\limsup_{n\to\infty}\sqrt[n]{u_{n-1}}=\lim_{n\to\infty} q_n$ again using the monotonicity of q_n [13].

Let us put $q=\lim_{n\to\infty} q_n$. Two cases naturally arise:

a) $q<1$

Since $p_n\downarrow 1-q>0$ and $u_n\leq q^n$ the chain is positive and geometrically ergodic with decay parameter $\rho_0=q$.

b) $q=1$

All three types of states can arise. In the transient case we have geometric ergodicity if $\lim_{n\to\infty}(p_n/p_{n+1})<1$ and the limit will be the decay parameter.

In the positive case there can be no geometric ergodicity by the above reasoning. In the null recurrent case we can infer that

$$\lim_{n\to\infty}\frac{P_{ij}^{(n)}}{P_{kl}^{(n)}}=\frac{u_{j-1}}{u_{l-1}}.$$

The argument uses a theorem in [2] to reduce the given statement to the simpler one,

$$\lim_{n\to\infty}\frac{P_{00}^{(n+1)}}{P_{00}^{(n)}}=1.$$

The latter follows from a theorem of Kingman-Orey [11, p. 22] in view of the monotonic character of the sequence $P_{00}^{(n)}$.

By repeating the elimination procedure in the proof of Szegö's theorem one can relax the condition on the q_n while still having a strong ratio-limit theorem.

2) *Unequal holding times*

In a lot of cases the holding time in a state is not a constant but depends randomly on the state. If all holding times are multiples of a common time unit we can take this unit as time unit in the success chain and put as many q_n equal to one as the holding time in the corresponding state requires.

In general however one has to use semi-Markov theory. It is our intention to develop this approach in a forthcoming publication [5], thus extending a few primary results along these lines obtained in [4].

References

1. AHLFORS, L. A. *Complex Analysis*. McGraw Hill, New York, 1966.
2. CHUNG, K. L. *Markov Chains with Stationary Transition Probabilities*. Springer-Verlag, Berlin, 1960.
3. DARROCH, J. N. and SENETA, E. On quasi-stationary distributions in absorbing discrete finite Markov chains. *J. Appl. Prob.*, **2**: 88–100; 1965.
4. EE SOO MEI, *Problems Relating to Semi-Markov Theory*. University of Malaya, 1971.
5. EE SOO MEI and TEUGELS, J. L. *On Semi-Markov Success Chains*. (forthcoming).
6. FELLER, W. *The Theory of Probability and its Applications*. J. Wiley, New York, 1957.
7. HARDY, G. H. *Divergent Series*. Clarendon Press, Oxford, 1949.
8. KARLIN, S. *A First Course in Stochastic Processes*. Academic Press, New York, 1966.
9. KEMENY, J. G., SNELL, J. L. and KNAPP, A. W. *Denumerable Markov Chains*. Van Nostrand, Princeton, New Jersey, 1966.
10. KENDALL, D. G. Geometric ergodicity and the theory of queues. ARROW, K., KARLIN, S. and SUPPES, P., eds. *Mathematical Methods in the Social Sciences*. Stanford University Press, California, 1960.
11. KINGMAN, J. F. C. *Regenerative Phenomena*. J. Wiley, London, 1972.
12. KNOPP, K. *Theorie und Anwendung der Unendlichen Reihen*. Springer-Verlag, Berlin, 1964.
13. POLYÀ, G. and SZEGÖ, G. *Aufgaben und Lehrsätze aus der Analysis*. Springer-Verlag, Berlin, 1925.
14. SENETA, E. and VERE-JONES, D. On quasi-stationary distributions in discrete-time Markov chain with a denumerable infinity of states. *J. Appl. Prob.*, **3**: 403–434, 1966.
15. TITCHMARSH, E. C. *The Theory of Functions*. Oxford University Press, London, 1939.
16. VERE-JONES, D. Geometric ergodicity in denumerable Markov chains. *Quart. J. Math.*, **13**: 7–28, 1962.

Chapter 19

PRACTICAL USE OF DECISION TREES

JACQUES FEDERWISCH

Summary

Numerous attractive formal studies on decision sequences have been published but most of them oversimplify. In concrete cases difficulties arise, such as the amount of mathematical computations involved and the correct selection among all possible alternatives.

The paper presents a complete analysis of a practical example, reviews the difficulties, and proposes simple operational methods of analysis, complementary to the abstract studies.

1. The Problem

The plant for a new line of production can be built by successively adding at most three units of identical capacity C. The alternatives are then: (1) to build capacity C, and enlarge later to 2C or 3C; (2) to build capacity 2C, and enlarge later to 3C; (3) to build capacity 3C.

Once a unit is built enlarging existing facilities will require two years. Funds for (3) are available but they, as well as subsequent returns, can be invested in a safe asset with guaranteed return of 7%. Again, invested funds may be reconverted into cash at 1% reconversion cost.

Cost and market data are as follows. Investment required for C, 2C, and 3C is (in Belgian francs) 80 m, 140 m, and 180 m, respectively. The plant can be dismantled and resold at any time, residual value varying between the following limits: resale of the respective capacities after 1 year yields 62 m, 110 m, and 140 m; after 10 years 9 m, 19 m, and 24 m.

According to the market study the probability of a 4-year lifetime is .55 (with second-degree probabilities .60 of market volume M and .40 of volume 2M); the probability of a 10-year lifetime is .45 (with second-degree probabilities .20 of M, .35 of 2M, and .45 of 3M). Furthermore competition will be severe so that any addition of supplementary capacity after the fifth year will be pointless. For different time periods the various market volumes have the probabilities indicated in table 1. Finally, given the annual profit curves for different sales volumes, table 2 states expected profits (in m francs) for the various probable market volumes (where bracketed amounts correspond to an excess of market volume over production capacity).

TABLE 1

Market volume \ Year		1	2	3	4	5	6	7	8	9	10
e_0	0 M					.55	.55	.55	.55	.55	.55
e_1	.5 M	.53			.33						.09
e_2	.75 M	.03									
e_3	1 M	.39	.42	.42	.31	.09	.09	.09	.09	.09	.16
e_4	1.5 M	.05	.11								.20
e_5	2 M		.27	.38	.16	.16	.16	.16	.16	.16	
e_6	2.25 M		.03								
e_7	2.5 M		.12								
e_8	3 M		.05	.20	.20	.20	.20	.20	.20	.20	

TABLE 2

Market Volume \ Capacity		C	2C	3C
e_0	0 M	−6.5	−12	−17
e_1	.5 M	4.5	−3.5	−10.5
e_2	.75 M	14	5	−4.5
e_3	1 M	22	14	4
e_4	1.5 M	25	30	22
e_5	2 M	(25)	37	37
e_6	2.25 M	(25)	(38)	42
e_7	2.5 M	(25)	(38)	46
e_8	3 M	(25)	(38)	50

2. Drawing and Computation of the Decision Tree

Drawing the decision tree without figures becomes complicated very soon. At the end of the second year 82 possible alternatives can be recorded and should be quantified. It is of course out of the question to carry on the investigation on that basis. Thanks to an exhaustive analysis of the problem, which has led to the definition of the above-mentioned numerical data, we will observe that certain simplifications of the drawing are acceptable without loss of relevance.

TABLE 3

Line of action	Investments for new plant		Corresponding investment with fixed returns
	Capacity	Amount invested	
a_0	—	—	180
a_1	C	80	100
a_2	2C	140	40
a_3	3C	180	—

TABLE 4

Line of action	State of nature	Industrial profits	Financial profits	Profits after 1 year
a_0		0	12.6	12.6
	e_1	4.5		11.5
a_1	e_2	14	7	21
	e_3	22		29
	e_4	25		32
	e_1	−3.5		−.7
a_2	e_2	5	2.8	7.8
	e_3	14		16.8
	e_4	30		32.8
	e_1	−10.5		−10.5
a_3	e_2	−4.5	0	−4.5
	e_3	4		4
	e_4	22		22

First Step

At the start (*beginning* of the first year), the four possible lines of action are summarized in table 3 (in m francs), while for each line of action expected profits at the *end* of the first year are tabulated in table 4.

Second Step

Depending on the possible lines of action, decisions at the *end* of the second step depend on *both* the state of the market at the end of the previous year and the financial possibilities. A computational example follows.

a) Combination (of lifetime and market volume) e_1

Derived line of action $a^1{}_{11}$: no enlarging, but profits are financially invested.

Derived line of action $a^1{}_{12}$: enlarging into a plant of 2C capacity.

required investments : 80 000 000
available cash : 4 500 000

funds to be cashed : 75 500 000
reconversion costs : 755 000

Total cost of investment : 80 755 000

Derived line of action $a^1{}_{13}$: enlarging into a plant of 3C capacity

required investments : 140 000 000
available cash : 4 500 000

funds to be cashed : 135 500 000
reconversion costs : 1 355 000

Total cost of investment : 141 355 000

This amount exceeds the company's own financial possibilities; this line of action *must* be discarded.

b) Combination e_2

Derived line of action $a^2{}_{11}$: no enlarging, but profits are financially invested.

Derived line of action $a^2{}_{12}$: enlarging to 2C.

Total cost of investment:

80 000 000+0.01 (80 000 000−21 000 000) = 80 590 000.

Derived line of action $a^2{}_{13}$: enlarging to 3C.
this line *must* be discarded.

c) Combination e_3

Derived line of action $a^3{}_{11}$: no enlarging, but profits are financially invested.

Derived line of action $a^3{}_{12}$: enlarging to 2C.

Total cost of investment:

80 000 000+0.01 (80 000 000−29 000 000) = 80 510 000;
this line *must* be discarded.

d) Combination e_4

Derived line of action $a^4{}_{11}$: no enlarging, but profits are financially invested.

Derived line of action $a^4{}_{12}$: enlarging to 2C.

Total cost of investment:

80 000 000+0.01 (80 000 000−32 000 000) = 80 480 000.

We are now able to draw up an overall table which allows not only to summarize the above conclusions but also to compute profits during the second year (table 5).
This simple *concrete* line of thought replaces the theoretical number of 82 arcs by only 29 arcs in practice, i.e. but 35.5%: this observation is essential.

Further Steps

One can proceed sequentially, using this practical analytic method of keeping but the actually possible orientations. In some instances a logical decision can be made at a given time, leading to a further offer of capacity exceeding potential sales volume. This applies to the lines of action:

$a_2 \to a^1{}_{21}$ corresponding to combination e_3 in the second year; excess capacity will be C;

a_3 corresponding to the combinations:
e_1 at the end of the 1st year: excess is C;
e_3 at the end of the 2nd year: excess is C (complementary);
e_5 at the end of the 2nd year: excess is C.

One must further consider the possibility of selling excess initial investment.

TABLE 5

Initial line of action	Combination first year	Available capital		Derived line of action	Investments		Combination second year	Profits of the 2nd year			Available financial amounts end 2nd year
		Main	Profits first year		Additional industrial	Total financial		Industrial	Financial	Total	
a_0	—	180.—	12.6	a_{00}	—	192.6	—	—	13.482	13.482	206.082
				a_{01}	80.674	111.926	—	—	7.835	7.835	119.761
				a_{02}	141.276	51.324	—	—	3.593	3.593	54.917
				a_{03}	181.676	10.924	—	—	0.765	0.765	11.689
a_1	e_1	100.—	11.5	$a^1{}_{11}$	—	111.5	e_3	22.—	7.805	29.805	141.305
							e_4	25.—	7.805	32.805	144.305
				$a^1{}_{12}$	80.755	30.745	e_3	22.—	2.152	24.152	54.897
							e_4	25.—	2.152	24.152	54.897
	e_2	100.—	21.—	$a^2{}_{11}$	—	121.—	e_6	25.—	8.470	33.470	154.470
				$a^2{}_{12}$	80.590	40.410	e_6	25.—	2.829	27.829	68.239
	e_3	100.—	29.—	$a^3{}_{11}$	—	129.—	e_5	25.—	9.030	34.030	163.030
							e_7	25.—	9.030	34.030	163.030
				$a^3{}_{12}$	80.510	48.490	e_5	25.—	3.394	28.394	76.884
							e_7	25.—	3.394	28.394	76.884
	e_4	100.—	32.—	$a^4{}_{11}$	—	132.—	e_8	25.—	9.240	34.240	166.240
				$a^4{}_{12}$	80.480	51.520	e_8	25.—	3.606	28.606	80.126

TABLE 5 (*continued*)

Initial line of action	Combination first year	Available capital		Derived line of action	Investments		Combination second year	Profits of the 2nd year			Available financial amounts end 2nd year
		Main	Profits first year		Additional industrial	Total financial		Industrial	Financial	Total	
a_2	e_1	40.—	−0.7	$a^1{}_{21}$	—	39.3	e_3	14.—	2.751	16.751	56.051
							e_4	30.—	2.751	32.751	72.051
	e_2	40.—	+7.8	$a^2{}_{21}$	—	47.8	e_6	38.—	3.346	41.346	120.146
	e_3	40.—	16.8	$a^3{}_{21}$	—	56.8	e_5	37.—	3.976	40.976	97.776
							e_7	38.—	3.976	41.976	98.776
	e_4	40.—	32.8	$a^4{}_{21}$	—	72.8	e_8	38.—	5.096	43.096	125.896
a_3	e_1	—	−10.5	—	—	−10.5	e_3	4.—	−1.050	2.950	−7.550
							e_4	22.—	−1.050	20.950	+10.450
	e_2	—	−4.5	—	—	−4.5	e_6	42.—	−0.450	41.550	+37.050
	e_3	—	4.5	—	—	4.—	e_5	37.—	0.280	37.280	41.280
							e_7	46.—	0.280	46.280	50.280
	e_4	—	22.—	—	—	22.—	e_8	50.—	1.540	51.540	73.540

[1] All values are expressed in millions of francs.
[2] The complementary industrial investments cannot generate profits before the beginning of the fourth year.

OPEN GRAPH OF DECISIONS

1. INITIAL LINE OF ACTION a_0

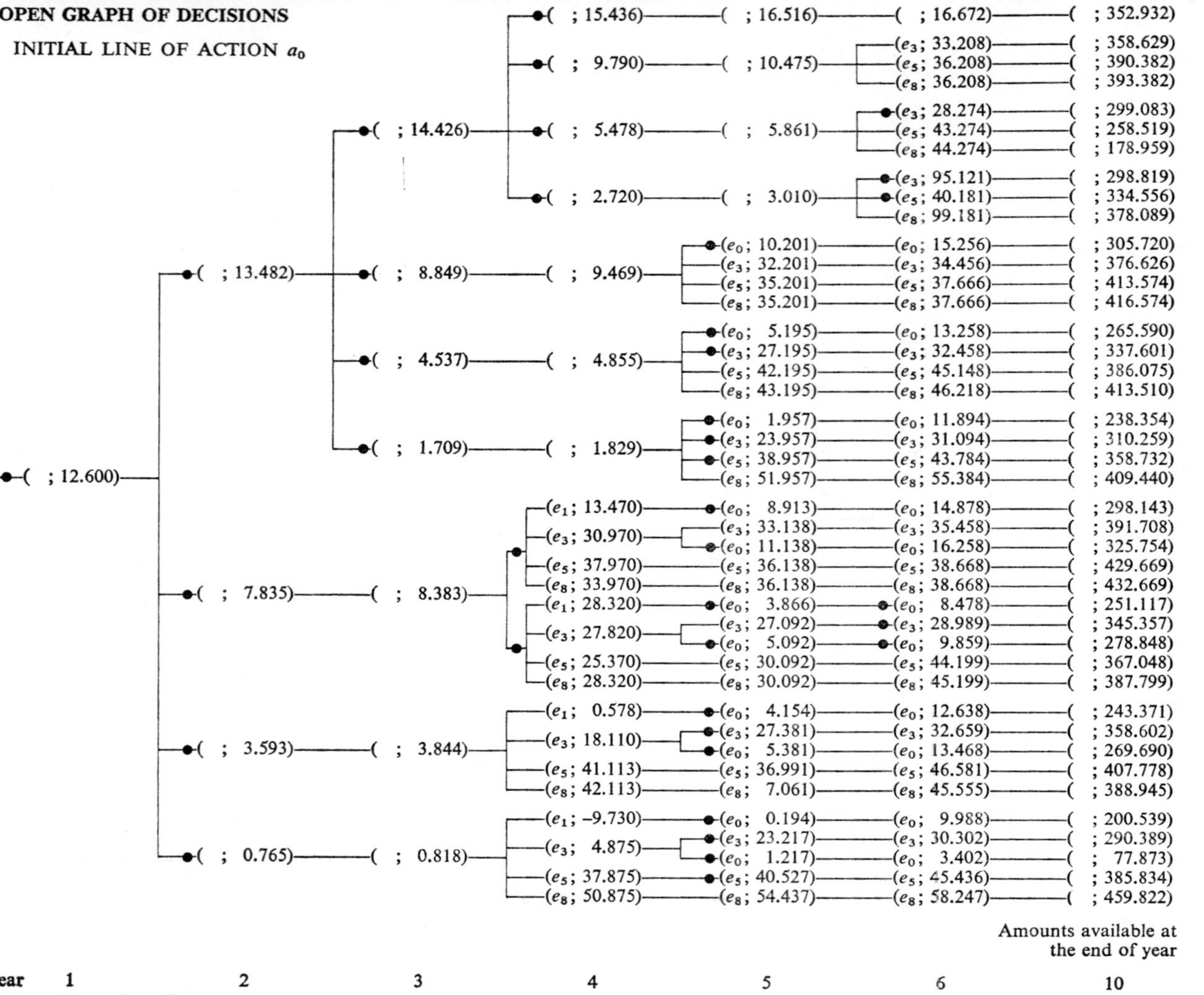

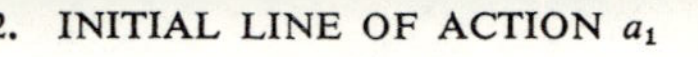

2. INITIAL LINE OF ACTION a_1

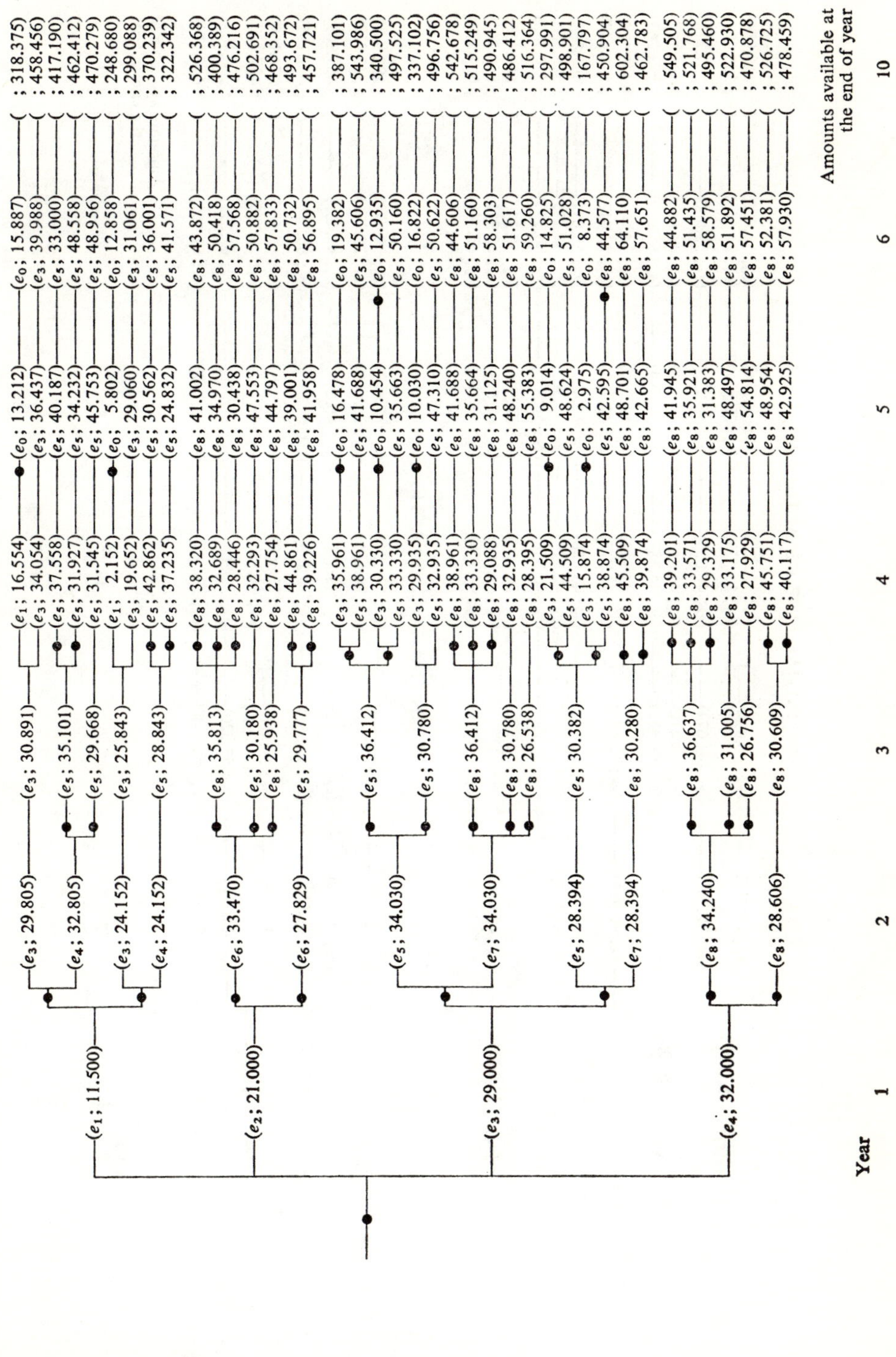

Year 1	2	3	4	5	6	10 (Amounts available at the end of year)
(e_1; 11.500)	(e_3; 29.805)	(e_3; 30.891)	(e_1; 16.554)	(e_0; 13.212)	(e_0; 15.887)	(; 318.375)
			(e_3; 34.054)	(e_3; 36.437)	(e_3; 39.988)	(; 458.456)
	(e_4; 32.805)	(e_5; 35.101)	(e_5; 37.558)	(e_5; 40.187)	(e_5; 33.000)	(; 417.190)
			(e_5; 31.927)	(e_5; 34.232)	(e_5; 48.558)	(; 462.412)
		(e_5; 29.668)	(e_5; 31.545)	(e_5; 45.753)	(e_5; 48.956)	(; 470.279)
	(e_3; 24.152)	(e_3; 25.843)	(e_1; 2.152)	(e_0; 5.802)	(e_0; 12.858)	(; 248.680)
			(e_3; 19.652)	(e_3; 29.060)	(e_3; 31.061)	(; 299.088)
	(e_4; 24.152)	(e_5; 28.843)	(e_5; 42.862)	(e_5; 30.562)	(e_5; 36.001)	(; 370.239)
			(e_5; 37.235)	(e_5; 24.832)	(e_5; 41.571)	(; 322.342)
(e_2; 21.000)	(e_6; 33.470)	(e_8; 35.813)	(e_8; 38.320)	(e_8; 41.002)	(e_8; 43.872)	(; 526.368)
			(e_8; 32.689)	(e_8; 34.970)	(e_8; 50.418)	(; 400.389)
			(e_8; 28.446)	(e_8; 30.438)	(e_8; 57.568)	(; 476.216)
		(e_5; 30.180)	(e_8; 32.293)	(e_8; 47.553)	(e_8; 50.882)	(; 502.691)
		(e_8; 25.938)	(e_8; 27.754)	(e_8; 44.797)	(e_8; 57.833)	(; 468.352)
	(e_6; 27.829)	(e_5; 29.777)	(e_8; 44.861)	(e_8; 39.001)	(e_8; 50.732)	(; 493.672)
			(e_8; 39.226)	(e_8; 41.958)	(e_8; 56.895)	(; 457.721)
(e_3; 29.000)	(e_5; 34.030)	(e_5; 36.412)	(e_3; 35.961)	(e_0; 16.478)	(e_0; 19.382)	(; 387.101)
			(e_5; 38.961)	(e_5; 41.688)	(e_5; 45.606)	(; 543.986)
			(e_3; 30.330)	(e_0; 10.454)	(e_0; 12.935)	(; 340.500)
			(e_5; 33.330)	(e_5; 35.663)	(e_5; 50.160)	(; 497.525)
		(e_5; 30.780)	(e_3; 29.935)	(e_0; 10.030)	(e_0; 16.822)	(; 337.102)
			(e_5; 32.935)	(e_5; 47.310)	(e_5; 50.622)	(; 496.756)
	(e_7; 34.030)	(e_8; 36.412)	(e_8; 38.961)	(e_8; 41.688)	(e_8; 44.606)	(; 542.678)
			(e_8; 33.330)	(e_8; 35.664)	(e_8; 51.160)	(; 515.249)
			(e_8; 29.088)	(e_8; 31.125)	(e_8; 58.303)	(; 490.945)
		(e_8; 30.780)	(e_8; 32.935)	(e_8; 48.240)	(e_8; 51.617)	(; 486.412)
		(e_8; 26.538)	(e_8; 28.395)	(e_8; 55.383)	(e_8; 59.260)	(; 516.364)
	(e_5; 28.394)	(e_5; 30.382)	(e_3; 21.509)	(e_0; 9.014)	(e_0; 14.825)	(; 297.991)
			(e_5; 44.509)	(e_5; 48.624)	(e_5; 51.028)	(; 498.901)
			(e_3; 15.874)	(e_0; 2.975)	(e_0; 8.373)	(; 167.797)
			(e_5; 38.874)	(e_5; 42.595)	(e_8; 44.577)	(; 450.904)
	(e_7; 28.394)	(e_8; 30.280)	(e_8; 45.509)	(e_8; 48.701)	(e_8; 64.110)	(; 602.304)
			(e_8; 39.874)	(e_8; 42.665)	(e_8; 57.651)	(; 462.783)
(e_4; 32.000)	(e_8; 34.240)	(e_8; 36.637)	(e_8; 39.201)	(e_8; 41.945)	(e_8; 44.882)	(; 549.505)
			(e_8; 33.571)	(e_8; 35.921)	(e_8; 51.435)	(; 521.768)
			(e_8; 29.329)	(e_8; 31.383)	(e_8; 58.579)	(; 495.460)
		(e_8; 31.005)	(e_8; 33.175)	(e_8; 48.497)	(e_8; 51.892)	(; 522.930)
		(e_8; 26.756)	(e_8; 27.929)	(e_8; 54.814)	(e_8; 57.451)	(; 470.878)
	(e_8; 28.606)	(e_8; 30.609)	(e_8; 45.751)	(e_8; 48.954)	(e_8; 52.381)	(; 526.725)
			(e_8; 40.117)	(e_8; 42.925)	(e_8; 57.930)	(; 478.459)

Fig. 1.2

3. INITIAL LINE OF ACTION a_2

Year 1	Year 2	Year 3	Year 4	Year 5	Year 6	Year 10 (Amounts available at the end of year)
(e_1; −0.700)	(e_3; 16.751)	(e_3; 25.924)	(e_1; 12.828)	(e_0; 9.226)	(e_0; 11.622)	(; 229.900)
			(e_3; 22.328)	(e_3; 29.891)	(e_3; 24.383)	(; 358.991)
	(e_4; 32.751)	(e_5; 42.074)	(e_5; 44.089)	(e_5; 48.075)	(e_5; 52.840)	(; 527.218)
(e_2; 7.800)	(e_6; 41.346)	(e_8; 46.410)	(e_8; 49.659)	(e_8; 53.135)	(e_8; 46.855)	(; 608.375)
		(e_8; 41.483)	(e_8; 44.387)	(e_8; 59.494)	(e_8; 63.659)	(; 572.263)
(e_3; 16.800)	(e_5; 40.976)	(e_5; 43.844)	(e_3; 23.913)	(e_0; 11.587)	(e_0; 15.401)	(; 308.624)
			(e_5; 46.913)	(e_5; 40.197)	(e_3; 53.011)	(; 530.642)
	(e_7; 41.976)	(e_8; 44.914)	(e_8; 48.058)	(e_8; 51.492)	(e_5; 55.097)	(; 567.897)
			(e_8; 42.424)	(e_8; 45.362)	(e_8; 60.579)	(; 516.556)
		(e_8; 39.288)	(e_8; 42.038)	(e_8; 56.981)	(e_8; 60.969)	(; 518.368)
(e_4; 32.800)	(e_8; 43.096)	(e_8; 46.813)	(e_8; 50.090)	(e_8; 53.596)	(e_8; 57.346)	(; 569.436)
			(e_8; 44.466)	(e_8; 47.579)	(e_8; 62.911)	(; 563.244)
		(e_8; 41.187)	(e_8; 44.769)	(e_8; 59.204)	(e_8; 63.348)	(; 604.370)

4. INITIAL LINE OF ACTION a_3

Year 1	Year 2	Year 3	Year 4	Year 5	Year 6	Year 10 (Amounts available at the end of year)
(e_1; −10.500)	(e_3; 2.950)	(e_3; 24.272)	(e_1; 11.131)	(e_0; 7.410)	(e_0; 9.678)	(; 193.961)
			(e_1; 28.631)	(e_3; 30.635)	(e_3; 32.779)	(; 332.032)
	(e_4; 20.950)	(e_5; 40.392)	(e_5; 43.219)	(e_5; 46.244)	(e_5; 49.481)	(; 459.909)
(e_2; −4.500)	(e_6; 41.550)	(e_8; 52.594)	(e_8; 56.289)	(e_8; 60.215)	(e_8; 60.215)	(; 489.251)
(e_3; 4.000)	(e_5; 37.280)	(e_5; 39.890)	(e_3; 21.362)	(e_0; 8.858)	(e_0; 12.487)	(; 250.240)
			(e_5; 44.362)	(e_5; 47.467)	(e_5; 50.790)	(; 437.623)
	(e_7; 46.280)	(e_8; 53.520)	(e_8; 57.266)	(e_8; 61.275)	(e_8; 61.275)	(; 510.489)
(e_4; 22.000)	(e_8; 51.540)	(e_8; 55.148)	(e_8; 59.008)	(e_8; 63.139)	(e_8; 67.559)	(; 636.520)

Fig. 1.3

Let us compute the residual value (in m francs) of the plant's units taken apart in each of the above instances: (1) capacity reduction from 2C to C, beginning of year 3, yields net residual values of 86 for 2C, 49 for C, and 37 for resale; (2) reduction from 3C to 2C, beginning of year 2, yields 140 for 3C, 110 for 2C, and 30 for resale; (3) reduction from 3C to 2C, beginning of year 3, yields 110 for 3C, 86 for 2C, and 24 for resale.

Consequently the complete table of logically possible decisions, corresponding to probable combinations of lifetimes and market volumes, is derived from the overall decision tree, which also indicates decision timing. In short, the study has been limited to 104 arcs for the 10-year period. Fig. 1 summarizes the graph.

3. Analysis of the Decision Tree

Starting from expected returns, previously computed, we draw the distribution curve of industrial profits compared with financial profits from a safe investment with a 7% annual rate of return.

The proportional distribution of profits takes the shape of a symmetric Gauss curve, whose amplitude varies between −275 059 and +283 588. Examination of this curve does not, however, allow to make any decision. We split the decision according to the nature of the initial potential decision and are led to a tentative order, viz. a_0, a_3, a_1, a_2, of profitability for the respective primary actions: a_2 appears to be most profitable.

In the previous examination the probabilities of the various market configurations were not taken into account. Table 6 shows the order, not of profits, but of overall returns, in a contingency table. (Compare: global return by investing capital at 7% is 352 932.)

In first approximation it appears that: (a) investing the capital with a fixed rate of return (a_0 financial market) is more profitable if the lifespan of the activity is 4 years; (b) as a rule a_0 is not, except in the above instance, the best solution and should be discarded.

In order to define a more accurate line of action expected profit *eP* and variance of profits *vP* are computed for each possible case. The ranking in order of increasing *eP* for states 1 through 8 is as follows: (1) a_3, a_2, a_1, a_0, a_p; (2) a_3, a_0, a_1, a_2, a_p; (3) a_3, a_0, a_p, a_2, a_1; (4) a_0, a_p, a_3, a_1, a_2; (5) a_0, a_p, a_1, a_3, a_2; (6) a_p, a_0, a_1, a_2, a_3; (7) a_p, a_0, a_3, a_1, a_2; (8) a_p, a_0, a_1, a_3, a_2. The *overall* ranking in order of increasing *eP* is then:

$$a_0, a_3, a_p, a_1, a_2. \tag{3.1}$$

TABLE 6

Market configuration								Initial lines of action						
REF	1	2	3	4	5 à 9	10	Probability	a_0			a_1		a_2	a_3
1	e_1	e_3	e_3	e_1	e_0	e_0	*0.33*	200.539 243.271 251.177 298.143	238.354 265.590		248.680 318.375		229.900	193.961
2	e_3	e_5	e_5	e_3	e_0	e_0	*0.22*	77.873 269.690 278.848 325.797	305.720		167.797 297.991 337.107	340.500 387.401	308.634	250.240
3	e_1	e_3	e_3	e_3	e_3	e_1	*0.09*	290.389 298.819 299.083 310.259	337.601 338.602 345.357 358.629	376.626 391.708	299.088 458.456		358.991	332.032
4	e_3	e_5	e_5	e_5	e_5	e_3	*0.05*	258.519 334.556 358.732 367.048	413.574 429.669		450.904 496.759 497.525	498.901 543.986	530.642	437.628
5	e_1	e_4	e_5	e_5	e_5	e_3	*0.11*	385.834 386.075 390.382 407.778			322.342 370.239 462.412	470.279 497.190	527.218	459.909
6	e_4	e_8	e_8	e_8	e_8	e_4	*0.05*	178.959 378.089 387.799			470.878 478.459 495.460 521.768	522.930 526.725 549.505	563.244 569.436 604.370	636.520
7	e_3	e_7	e_8	e_8	e_8	e_4	*0.12*	388.945 393.382 409.440 413.510			462.783 486.412 490.945 515.240	516.364 542.678 602.304	516.556 518.368 567.897	510.489
8	e_2	e_0	e_8	e_8	e_8	e_4	*0.03*	416.574 432.669 459.822			400.389 457.721 468.352 476.216	493.672 502.691 526.368	572.263 608.375	489.251

Similarly one obtains the *overall* ranking in order of decreasing vP:

(3.2) $$a_3, a_2, a_1, a_0, a_p.$$

One restricts choice to actions a_i, a_j *efficient* in terms of *both* moments eP, vP (cf. Markowitz [1] and Van Moeseke [2, 3]), i.e. a_i is efficient if, for

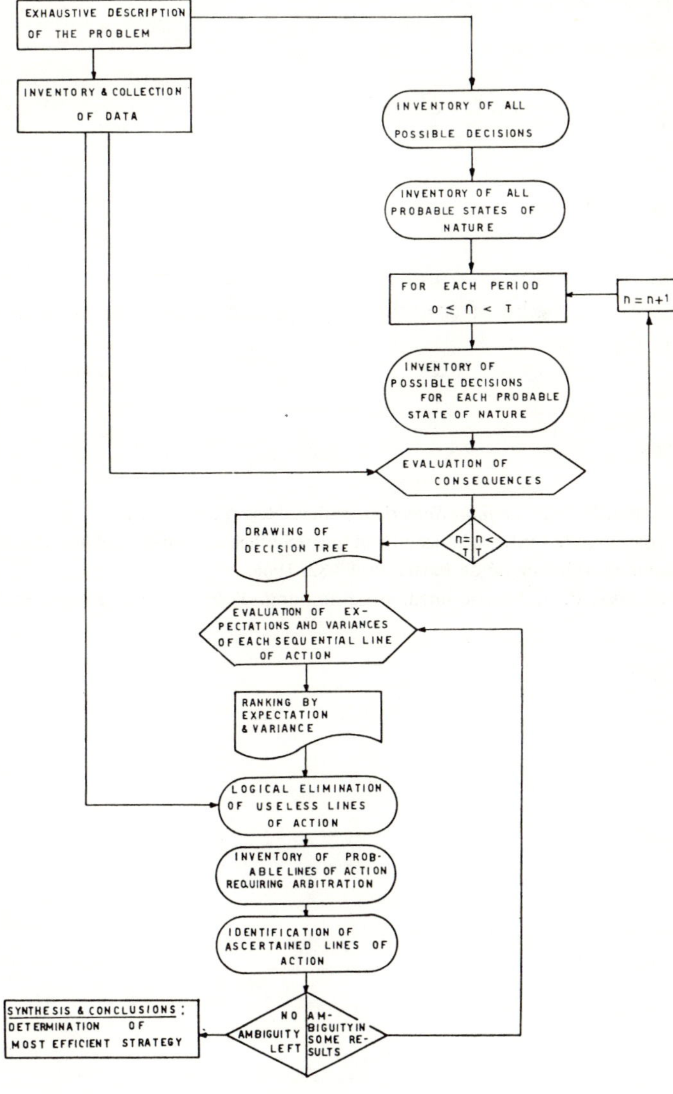

Fig. 2

all j, $eP_i \leq eP_j$ implies $vP_i \leq vP_j$ *and* $vP_j \leq vP_i$ implies $eP_j \leq eP_i$. One readily checks from (3.1), (3.2) that a_1, a_2, a_p are efficient.

Further choice may be based on a postulated expected-utility function eU where $U = a + bP + cP^2$ and b, c pertain to the decision-maker's risk psychology. One has

$$eU = a + beP + ceP^2$$

or, as $eP^2 = vP + e^2 P$,

$$eU = a + beP + c(vP + e^2 P).$$

4. Schematic Outline

In conclusion of the complete study of a practical case, simplified somewhat for the sake of computation, we summarize the methodological scheme in fig. 2.

References

1. MARKOWITZ, H. M. *Portfolio Selection.* Wiley, New York, 1959.
2. VAN MOESEKE, P. Ordre d'efficacité et portefeuilles efficaces. *Cahiers du Séminaire d'Econométrie* (Université de Paris), **9**: 67–82, 1966.
3. VAN MOESEKE, P. Stochastic linear programming. *Yale Economic Essays*, **5**: 196–254, 1965.

CHAPTER 20

BACTERIOLOGY LABORATORY COSTS: A UNIVERSALLY APPLICABLE MODEL

KATRIEL WILD

1. Introduction

Measures of effectiveness for decision making purposes are extremely difficult to obtain in the hospital environment. This is especially true in those areas where cost is the primary measure by which comparisons and evaluations are made. These costs, due to the dynamic nature of inflation and technological change, are cumbersome to identify even if one chooses to estimate costs at a particular instant of time. In order for costs to be useful for decision making, the development of a system of cost information is of primary importance. Moreover, due to the great expense involved in estimating costs, it is desirable to construct cost models which are generally adaptable to any size or type of hospital.

A significant contributor to hospital costs is the Department of Clinical Laboratories [5]. A study by Delon [3] explored cost methodologies for the Clinical Laboratories in a single hospital, citing the need for an updating device to counterbalance frequent fluctuations in laboratory costs. These cost fluctuations, characteristic of a dynamic system, are further complicated by several probabilistic components imbedded in particular sections of Clinical Laboratories. The most complex of these sections is the bacteriology laboratory.

Studies pertaining to costing in bacteriology have been conducted [1, 2]; however, they have been concerned with costing at a single instant of time. They have not considered the stochastic properties of many of the individual tests and for most hospitals are too expensive to reproduce and apply.

A bacteriology specimen, after being received at the laboratory, is inoculated onto media plates and is immediately incubated. Following incubation, the treatment given to the specimen depends on the technician's findings. This differs from most other clinical tests in that a bacteriology specimen may follow any one of numerous paths given the finding of the preceding stage. In other clinical sections a specimen usually follows a prescribed path which leads to some accepted conclusion. However, in bacteriology tests, the probability of stage j (where $j = 1, 2, \ldots, n$) occurring is conditional upon the outcome at stage $j-1$. Therefore, the costs of bacteriology tests vary depending upon the number of stages of treatment. To illustrate, consider a specimen which is found to be sterile; it is discarded immediately. Its processing cost is considerably smaller than that of a specimen which requires a long sequence of various laboratory operations until a result is obtained.

The method proposed here not only recognizes the basic difference between bacteriology and other clinical laboratory areas, i.e. that it is a stochastic system, but also develops a model which can be applied to most hospitals with very little modification.

2. Analysis of Laboratory Operations

As previously mentioned, the first process a specimen undergoes is its inoculation onto media plates; this is the "planting" stage. The bacteriology laboratory processes approximately fifty different kinds of specimens, yet these can be grouped according to their planting stage, since a given planting procedure may be common to several kinds of specimens. The entire process a specimen goes through is viewed as a branching process, as depicted in fig. 1.

In general, specimens which have a similar planting procedure also have the same branching process. This fact allows grouping of tests into fifteen test groups as follows: Blood, Bronchial, Aspirate, Cervix, Ear, Eye, Spinal Fluid, Sputum, Sterility Check, Stomach Aspirate, Stool, Throat, Tissue, Umbilicus, Urine and Wound. A test is defined as the process starting when the specimen is received at the Laboratory and ends when a final report is issued.

A detailed flow chart was prepared for each test, completely depicting the branching process through which a test may go. Flow charting was done at three different hospitals: a general hospital, a pediatric hospital and an obstetric general hospital. For each given test the flow charts from the three hospitals were compared. Few differences were found among the flow charts except for certain specific processes. These processes were peculiar to each hospital and differences were related to the nature of the patients; *i.e.* adults in one hospital, children in the second and mainly women in the third. A master flow chart was prepared depicting the processes in the three

In this illustration, a specimen may either follow path A - B - C - D, or path A - B - E - F - G - H - I or path A - B - E - F - G - J - K.

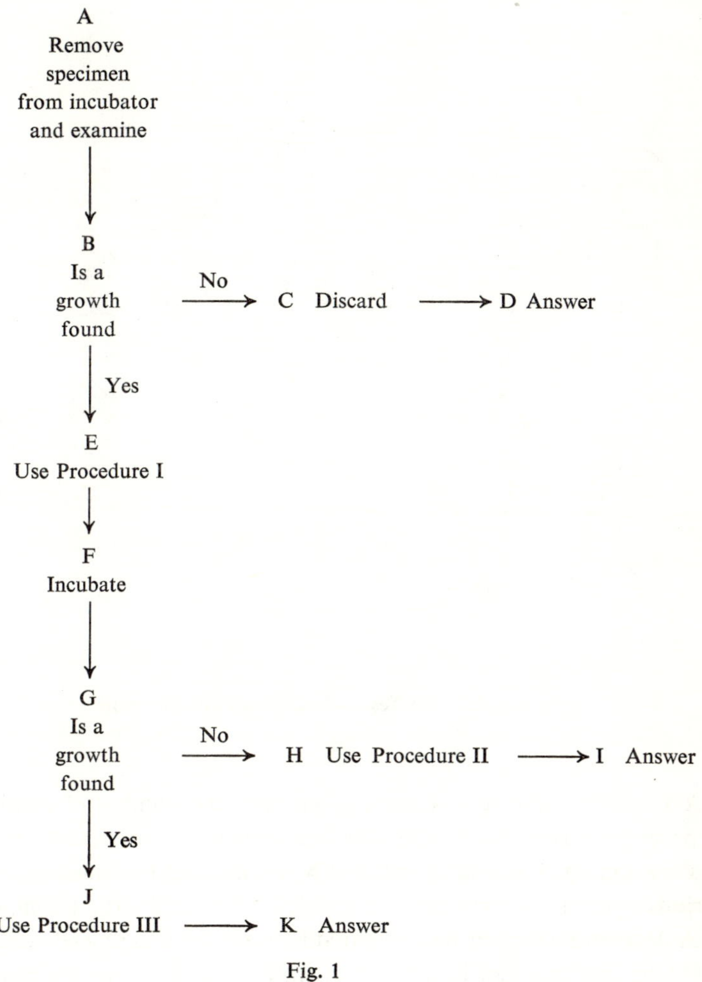

Fig. 1

hospitals such that by deleting certain steps the bacteriology laboratory at any one of the hospitals could be adapted to the master chart.

Fig. 2 describes a branching process for the Urine test where the diamond-shaped cells indicate decision points. Since two paths stem from each decision point the number of different possible paths for the specimen to follow is 2^n, where $n =$ number of decision points. Since in fig. 2, $n = 9$, the number of possible paths is $2^9 = 512$. The large number of possible paths (some tests have greater n-values) requires a simplified system of flow charts where the number of possible paths is more manageable.

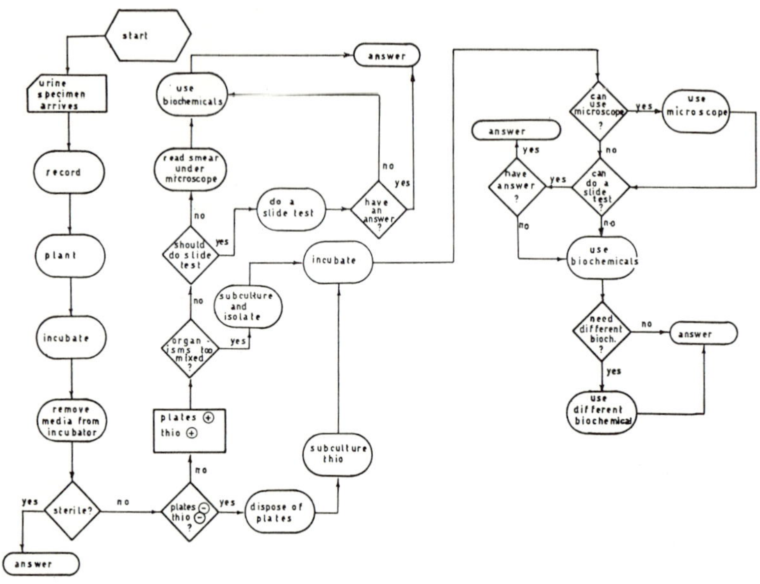

Fig. 2. Urine Test—Procedure Flow Diagram.

A simplified scheme is suggested whereby the number of possible paths or procedures is reduced to 20. The breakdown of each test into procedures, as shown in fig. 3, was accomplished by noting that the independence of the various operations permitted the simple addition of certain components of cost. Moreover, certain operations were so insignificant costwise that they could be incorporated into other operations specified in the chart. Using this chart, it is easy to distinguish among the different costs pertaining to each test. For example, Procedure 20 of the Urine test is by far less expensive than Procedure 10. This is so because Procedure 20 indicates "No Growth", namely no further processing is required.

The cost of each procedure can be determined by properly allocating salary, overhead and materials cost. Although management at this point knows the cost per individual procedure, it still lacks information as to the

Operation

Procedure	Plates $\ominus$ Thio $\oplus$	Sub Thio and Re-incubate	Plates $\oplus$ Thio $\oplus$	Slide test	Organisms too mixed	Subculture and isolate	Re-incubate	Biochemicals	Questionnable growth in thio	Re-incubate thio	Take out and sub. thio	Re-incubate thio	Use microscope
1.			*	*	*	*	*						
2.			*	*	*	*	*						*
3.			*	*									
4.			*	*				*					
5.			*	*				*					*
6.			*	*	*	*	*	*					
7.			*	*	*	*	*	*					*
8.			*		*	*	*	*					
9.			*		*	*	*	*					*
10.	*	*		*				*					
11.	*	*		*				*					*
12.	*							*	*	*	*	*	
13.	*	*						*	*	*	*	*	*
14.	*			*					*	*	*	*	
15.	*			*					*	*	*	*	*
16.	*	*		*									
17.	*	*		*									*
18.													
19.	*	*											*
20.						NO GROWTH							

Fig. 3. Procedures for Urine Test.

frequency of each procedure and, therefore, cannot estimate the cost of a specific test. The cost of Procedure 20 is much lower than that of Procedure 10, yet the former is by far more frequent than the latter. Thus frequencies of the various procedures are needed in order to obtain an overall estimate of the cost of a Urine test. This frequency information can be easily obtained in most laboratories.

Laboratory technicians keep a record of operations performed while processing each specimen. A 90-day sample was extracted from the laboratory record file and all test descriptions were matched against the procedure table (fig. 3). Tests were classified according to the procedure followed. From the total number of tests recorded, the frequency of each test procedure can be calculated in terms of probabilities. Therefore, if the monthly number of Urine tests was 500 and the frequency of Procedure 20 was 300, one may conclude that there is a probability of .6 that Procedure 20 will be applied to a specimen received in the laboratory. The 90-day sample taken from the technician's records for the Urine test gave these probability figures: Procedure 20 = .599, Procedure 8 = .125, Procedure 9 = .086 and Procedures 1 through 7 and 10 through 19, all combined = .190. All tests were broken down into procedures in the above manner, and probabilities were established pertaining to these test procedures.

3. Cost Allocation

Cost elements measured in the laboratory are primarily salary and material costs. The first is determined by the time required by each test and the second is based on the cost of the actual materials used in the test. Overhead and equipment depreciation costs are also considered. These are based on a unit-value structure which is developed in this study and which is related to a test's time consumption.

Costs frequently change and, therefore, the cost model developed makes provision for simple and efficient updating of cost data. Items such as material cost or salaries are subject to frequent fluctuations and the cost model has to adapt. The time required for a particular test may also change as procedures change; for example, a decision may be made to include a pathologist's inspection of a specimen. Cost of a test is obviously affected by such a decision requiring the addition of a time element. More effective methods leading to a reduction in the number of employees or a change in employee status (from full to part time) must also be considered in programs for updating costs in the laboratory.

3.1. Time Study and Work Sampling

In order to account for time required to perform an individual test, a time study and work sampling were conducted. Time study observations were taken over a four-month period at the three hospitals. It was found difficult to time individual operations as originally defined since these were actually a combination of sub-operations, not always performed in the same order. The sub-operations were then identified and timed. A sufficient number of time study observations was taken in order to accurately represent average time for a particular sub-operation. Student's *t* test was used to determine the number of observations required in order to arrive at the desired accuracy. The time elements derived were to be used as "building blocks" in establishing the time for a given test procedure.

Consider an example, where a test procedure consists of these items:

(a) Plates⊕Thio⊕
(b) Slide test
(c) Organisms too mixed
(d) Subculture and isolate
(e) Re-incubate
(f) Use microscope.

Items (a) and (c) constitute outcomes rather than operations and therefore, no time elements are associated with them. The other items are listed in the following table, along with the '*Record*', '*Plant*' and '*Incubate*' elements which are common to all procedures.

Description of time elements (building blocks)	Item	Time element
Record		.45 min.
Plant		3.33
Incubate		.06
Slide test	b	5.75
Arrange plates and tubes	d	.30
Read plates and tubes	d	4.73
Re-incubate	e	1.45
Use microscope	f	1.82
Total time		17.89 min.

Note that item (d) consists of two time elements. The first of these elements (arrange plates and tubes) can be applied also as part of several other operations, e.g., "biochemicals". Thus the total time consists of individual building blocks, some of which are further divided into secondary building blocks. A table of building blocks was established for all suboperations, which, in varying compositions, can constitute many different operations. See fig. 4.

The total test time derived in the previous example is 17.89 minutes, but this relates only to the direct work involved with the particular specimen. Realizing that laboratory workers are paid for their entire working day, it is necessary to account for any time involved in non-direct laboratory activities and incorporate this into the time required for each test. Non-direct time was calculated using work sampling. A small preliminary sample of observations was taken in order to determine the number of observations required to attain the desired accuracy of the final result. The results of the preliminary sampling done in this study indicated 17% non-direct time, and it was decided that 15% precision with 95% confidence would be acceptable. The following formula was used to calculate the number of observations necessary for a prescribed precision:

$$n = \frac{Z^2(1-p)}{S^2 p},$$

where

S = precision expressed as a decimal
p = percentage expressed as a decimal
n = number of observations
Z = the normal deviate for 95% confidence.

The number of observations required was:

$$n = \frac{(3.841)(.83)}{(.0225)(.17)} = 833.$$

Work sampling revealed 15.48% of non-direct time based on 950 actual observations. In order to determine the actual precision of the work sampling the following equation was used:

$$S = \frac{Z^2(1-p)}{np} = \frac{(3.841)(.8452)}{(950)(.1548)} = .148.$$

This result was within the specified precision. The work sampling study was done at different hospitals and no significant difference was found between the results obtained at individual hospitals.

Elements	Blood	Spinal Fluid	Cervix	Tissue	Wound	Throat	Urine	Umbilicus	Sputum	Stomach Aspirate	Stool	Bronchial Aspirate	Ear	Eye	Sterility Check
1. RC	D.A.	.825	.825	.710	1.10	.820	.450	1.10	.825	.825	.825	.825	.825	.825	.825
2. P	D.A.	4.20	3.50	3.83	4.18	2.29	3.33	4.18	3.90	3.10	4.20	4.75	3.93	3.93	1.84
3. I	D.A.	.54	.54	.54	.30	.29	.06	.30	.54	.54	.54	.54	.54	.54	.54
**4. GS	D.A.	.85	.85	.85	.85	.85	.85	D.A.	.85	.85	D.A.	.85	.85	.85	D.A.
5. Rsm	D.A.	2.24	1.97	2.30	2.30	1.70	1.82	D.A.	2.24	2.24	D.A.	2.30	2.30	2.30	D.A.
**6. A-p-t	D.A.	.29	.31	.36	.36	.36	.30	.36	.36	.32	.32	.36	.36	.36	.50
7. R st	D.A.	1.50	D.A.	1.50	1.50	1.50	1.50	1.50	1.50	1.50	1.50	1.50	1.50	1.50	1.50
*8. Rp. t-(b)	D.A.	D.A.	3.89	2.94	1.59	2.48	3.21	1.59	D.A.	2.37	D.A.	D.A.	2.94	D.A.	D.A.
*9. Rp. t-st(b)	D.A.	D.A.	5.75	5.75	5.75	5.75	5.75	5.75	5.75	5.75	5.75	5.75	5.75	5.75	5.75
10. Rp. t-ri	D.A.	1.25	1.80	2.13	2.60	2.27	1.45	1.45	1.86	1.39	2.60	2.62	2.13	2.13	D.A.
*11. Rp. t-sn(b)	D.A.	1.69	1.69	1.69	1.69	1.69	1.69	1.69	1.69	1.69	1.69	1.69	1.69	1.69	D.A.
12. Rp. t-sg	D.A.	4.90	4.90	4.90	4.90	4.90	D.A.	4.90	4.90	4.90	7.15	4.90	4.90	4.90	D.A.
*13. Rp. t-sm(b)	D.A.	3.32	3.32	3.10	3.20	4.18	D.A.	2.90	3.80	3.80	4.00	3.80	3.10	3.10	D.A.
14. Rp. t-i	D.A.	4.73	4.73	4.73	4.73	4.73	4.73	4.73	4.73	4.73	4.73	4.73	4.73	4.73	D.A.
15. SUb	D.A.	2.84	2.26	3.60	1.52	2.84	2.51	1.52	3.00	1.65	3.00	3.70	3.60	3.60	2.67
16. Rb	D.A.	1.67	3.75	1.85	2.60	2.39	1.56	2.60	2.00	1.20	3.70	0.65	1.85	1.85	2.00
17. Rs	D.A.	2.60	.39	1.10	2.60	1.65	.25	2.60	2.80	1.16	2.80	3.75	1.10	1.10	1.85
18. Growth	36.98														
19. No Growth	17.79														

* Wherever (b) appears add elements 15, 16 if biochemicals were added.
** Variable Time Element.
D.A. – No time figure assigned to these elements.

Fig. 4. Time Elements.

3.2. *The Relative Unit Value*

For allocating salary, overhead and depreciation costs according to time spent for each test, the Relative Unit Value (RUV) method is convenient. Allocating costs on the basis of time requires the manipulation of small time figures achieving accuracy neither warranted nor useful. The RUV is a common unit for which a base is established according to some arbitrary time element. In this case, the base for the RUV method was chosen as 15 minutes. Therefore, a test or procedure which consumes 15 minutes is considered as 1 RUV. A dollar value can be assigned to each RUV to simplify costing. If the amount assigned to 1 RUV = \$3.00, then a test which takes 18 minutes will cost \$3.60.

At this point, the results of the time study, work sampling and the RUV method can be combined. Using the example given earlier, where the direct time involved in a test procedure was 17.89 minutes and non-direct time was 15.48% of total time, total time required for a particular test procedure can be computed as follows:

$$t = d + \frac{d \times n}{100 - n},$$

where d = direct time; n = % idle time; t = total time.
Using the data from this example:

$$t = 17.89 + \frac{(17.89) \times (15.48)}{100 - 15.48} = 21.16 \text{ min}$$

21.16/15.00 = 1.41 RUV.

This RUV requirement serves as a basis for the allocation of salary, materials, overhead, and depreciation costs. The allocation of each component cost is discussed and the combined cost is provided in the next section, making use of the RUV values for the various test procedures.

3.3. *Salary Costs*

The allocation of salary costs to individual procedures and tests is done according to the number of RUV used. Direct salaries, i.e. those of the laboratory employees; and indirect salaries, i.e. clerical and managerial, are considered according to the percentage of time spent by each employee at the laboratory (fig. 5).

Employee	(a) annual full-time salary	(b) % allocated to Bacteriology	(a) × (b) salary allocated to Bacteriology
Technologist	$ 7,987.20	100.0%	$ 7,987.20
Lab. Aide	5,283.20	100.0	5,283.20
Lab. Aide	5,000.00	24.0	1,200.00
Technician	7,820.80	100.0	7,820.80
Technician	7,820.80	100.0	7,820.80
Technologist	7,696.00	100.0	7,696.00
Technician	6,780.80	100.0	6,780.80
Technologist	8,819.20	100.0	8,819.20
Technologist	7,280.00	38.7	2,800.00
Lab. Supervisor	10,982.40	100.0	10,982.40
Clerk	4,000.00	60.0	2,400.00
Ass't. Director	14,330.00	15.0	2,150.00
Pathologist*	26,666.00	15.0	4,000.00
Total			$75,840.40
Employee salaries related to materials cost:			
Media maker	5,927.50	80.0	4,742.00
Glassware washer	4,430.00	50.0	2,215.00
Total			6,957.00

* The pathologist's task is supervision and inspection of tests, and although he may not inspect all tests at one time his attention is given to all tests over the year.

Fig. 5. Bacteriology Salaries.

The salary cost assigned to a test procedure is calculated in the following manner. Let

N_i = number of employees in ith percentage category;
T = total annual salary cost;
P_i = percentage of time ith employee works in bacteriology;
Y = total yearly man hours;

$(\mathrm{RUV})_{ij}$ = relative unit value of test i, procedure j;
V = total yearly number of RUV;
D = dollar amount for labor assigned to 1 RUV;
S_{ij} = total salary cost for test i, procedure j;
r = time in hours for 1 RUV.

Then

$$S_{ij} = (\mathrm{RUV})_{ij} \times D,$$

where

$$D = T/V$$

and

$$V = [Y/r] \times [\sum_{i=1}^{n} N_i P_i];$$

therefore

$$S_{ij} = \frac{(\mathrm{RUV})_{ij}\, Tr}{Y \sum_{i=1}^{n} N_i P_i}.$$

To illustrate this calculation S_{ij} is to be determined for Urine test number 7, procedure number 2.

$(\mathrm{RUV})_{7,2}$ = (21.16 minutes)/(15 minutes) = 1.41 RUV
T = \$75,840.40 (see fig. 5)
r = 0.25 hours
Y = 2080 hours

$N_1 = 8$	$P_1 = 1.00$ (100%)	$N_1 P_1 = 8.000$
$N_2 = 1$	$P_2 = .240$	$N_2 P_2 = .240$
$N_3 = 1$	$P_3 = .387$	$N_3 P_3 = .387$
$N_4 = 1$	$P_4 = .600$	$N_4 P_4 = .600$
$N_5 = 2$	$P_5 = .150$	$N_5 P_5 = .300$

$$\sum_{i=1}^{5} N_i P_i = 9.527$$

Therefore

$$S_{ij} = \frac{(1.41)(\$75{,}840.40)\,(0.25 \text{ hours})}{(2080 \text{ hours})\,(9.527)} = \$1.35\,.$$

3.4. *Cost of Materials*

Materials used at the laboratory can be divided into two categories:

(a) materials prepared in the laboratory;
(b) materials purchased ready for use.

The cost allocation is done in the following manner. The cost of materials which are always purchased is allocated to each test procedure according to the kinds of materials used for this procedure. For example, if the price of a certain media plate is $0.25, this will be the cost allocated to the procedure in which the particular plate is used. The cost of materials which are prepared in the laboratory is allocated to each procedure according to the RUV of that procedure. A total cost for laboratory-prepared materials is calculated for a period of one year because certain types of raw materials and supplies are purchased only once a year. This total cost consists of

(a) raw materials cost;
(b) supplies cost;
(c) media-maker's salary;
(d) glassware washer's slary.

The total cost mentioned divided by the total yearly number of RUV's gives the cost of laboratory-prepared materials assigned to one RUV. Laboratory policies change frequently and some materials, which are usually laboratory-prepared, are temporarily purchased. Later the policy may change again and the particular materials would again be laboratory-prepared. The RUV cost allocation is, therefore, applied to the materials which are always laboratory-prepared and to those which are subject to policy change. The materials cost assigned to a test procedure is calculated in the following manner. Let

M_i = cost of purchased material i;
R_{ij} = total cost of prepared materials allocated to test i, procedure j;
L = total annual cost of all laboratory-prepared materials;
A = annual cost of raw materials;
E = annual cost of supplies;
K = media-maker's salary;
G = glassware washer's salary;
V = total yearly number of RUV's;
C_{ij} = total material cost for test i, procedure j.

Then

$$C_{ij} = R_{ij} + \sum_{i=1}^{n} M_i,$$

where

$$R_{ij} = [(\mathrm{RUV})_{ij}] \times [L/V],$$

$$L = A+E+K+G.$$

Therefore

$$C_{ij} = (\mathrm{RUV})_{ij} \times (A+E+K+G)/V + \sum_{i=1}^{n} M_i.$$

To illustrate this calculation: determine C_{ij} for the Urine test (No. 7), procedure 2.

$$(\mathrm{RUV})_{7,2} = 1.41$$

$A =$ \$6350.00 $\qquad$ $K =$ \$4,742.00
$E =$ 9063.00 $\qquad$ $G =$ 2,215.00
$V =$ 79,264.64 RUV's

$M_1 =$ \$.17 $\qquad$ $M_4 =$ \$.18
$M_2 =$ .17 $\qquad$ M_5 through $M_{16} = 12 \times$ (\$.02) $=$ \$.24
$M_3 =$ .17

$$\sum_{i=1}^{n} M_i = \$.93.$$

Therefore C_{ij} equals (in \$):

$$1.41 \times \frac{6{,}350+9{,}063+4{,}742+2{,}215}{79{,}264.64} + .93 = 1.33.$$

3.5. Overhead and Depreciation Cost

A certain amount of overhead and depreciation cost is allocated by the administration to the bacteriology laboratory and that cost, in turn, is distributed among the different tests performed at the laboratory. Overhead and depreciation costs are allocated according to the RUV method to assure equitable cost allocation. It may be argued that equipment depreciation cost should be allocated only to the tests for which the equipment is utilized.

However, since most equipment is used for all tests, the RUV method of allocating depreciation cost is suitable.

The overhead and depreciation cost assigned to a test procedure is calculated as follows: let

Q = annual overhead cost allocated to bacteriology;
U = annual depreciation cost allocated to bacteriology;
$(\text{RUV})_{ij}$ = RUV of test i, procedure j
V = total yearly number of RUV's;
B_{ij} = total overhead and depreciation cost allocated to test i, procedure j.

Then

$$B_{ij} = [(\text{RUV})_{ij}] \times [(Q+U)/V] .$$

To illustrate this calculation for test 7 (Urine), procedure 2,

$(\text{RUV})_{ij}$ = 1.41
Q = \$26,756.00
U = 1,682.00
V = 79,264.64 RUV's.

Therefore

$$B_{7,2} = (1.41) \times (\$26{,}756.00 + \$1{,}682.00)/(79{,}264.64) = \$.50.$$

3.6. *Total Cost of Test*

To find the full cost of performing test i, procedure j, the three cost components can be combined to yield

$$Z_{ij} = S_{ij} + C_{ij} + B_{ij} .$$

For the cost of test 7, procedure 2,

$$Z_{7,2} = \$1.35 + \$1.330 + \$0.50 = \$3.18 .$$

Having established the cost of procedure j of test i, knowledge of the probability for each procedure becomes useful for budgeting purposes. The average annual number of specimens of test i is known and can be used to obtain an estimate of the number of these specimens in the coming year. For example, if the probability of procedure j of test i is .07 and it is predicted that there will be 6,000 specimens of this test annually during the next year, then the number of specimens of test i following procedure j will be $6{,}000 \times .07 = 42$. By knowing the cost of procedure j of test i, the total budget for the coming year can be estimated.

A computer program was designed to handle the entire cost allocation, considering the various procedure probabilities and all cost components. The program was written so as to enable a simplified updating of laboratory data, such as materials costs, employees salaries, overhead amount allocated to bacteriology and other factors affecting cost such as the addition or elimination of operations for a particular test procedure.

Recognizing the fact that procedure probabilities are subject to change, a control method was devised for the periodic updating of procedure probabilities. Updated procedure probabilities along with the cost of each test procedure provide a sound basis for management decisions concerning budgeting and pricing. Forecasting is aided by this cost methodology and, therefore, long range planning is easier.

4. Summary and Conclusions

This study develops a model for calculating and updating bacteriology laboratory costs in a manner which can be easily adapted to most hospital facilities. The system presented is geared to overcome the dynamic inaccuracies and high expense of cost analyses which typically rely on initial data and have little or no provision for maintaining and updating these costs. The bacteriology laboratory was chosen as the first area in which multiple-applicability models were applied because bacteriology tests are well-defined and, generally, are similar regardless of the hospital. Even though specific procedures differ from hospital to hospital, the overall methods for a given test are similar and are complicated only by differences in the stochastic properties associated with that test.

It is important to note that ownership of a computer is not required to implement this costing system. All data for maintenance of the system are routinely available already. The initial investment of time required is limited to comparison of the work flow at the particular hospital with the master flow chart already available. Computer time required for each recalculation of cost is less than two minutes on an IBM 360 with the hospital being provided with a printout of costs for each test and procedure combination. The program is written in Fortran IV and may be used in almost any time-sharing system or service center.

The ability to obtain these costs in an accurate, timely, and well-structured form provides for many potential side benefits. The effects of automation or other procedural changes can be evaluated prior to making actual changes. Merging of facilities for several hospitals can be simulated and evaluated using the information provided by this system. Models can be developed for staffing the bacteriology laboratory, for estimating materials required, and for budgeting. If automation is introduced instead of the manual reporting system currently used to compile test results, better reports pertaining to a patient specimen could be provided to the physician and, in addition, could be stored for research purposes at almost negligible cost.

Potential uses of costing systems such as this one are too numerous to fully delineate, but it is certain that good management and pricing is difficult, if not impossible, without accurate cost data. This first of a series of multiple-applicability cost models is a step towards providing these data at an expense within the reach of most hospitals.

References

1. Cheatle, E. L. Computer tabulates bacteriology reports. *Hospitals, I.A.H.A.*, January 1964.
2. Chicago Hospital Council, *Clinical Laboratory Study*. 1965.
3. Delon, G. L. *Cost Methodologies in the Pathology Laboratory*. University of Pittsburgh, 1968.
4. Heiland, R. E. and Richardson, W. J. *Work Sampling*. McGraw Hill, New York, 1957.
5. Metropolitan Teaching, *Comparative Financial and Statistical Information*. Blue Cross of Western Pennsylvania, Pittsburgh, Pennsylvania, 1969.

INDEX OF REFERENCES

SUBJECT INDEX